SpringerBriefs in Applied Sciences and Technology

SpringerBriefs present concise summaries of cutting-edge research and practical applications across a wide spectrum of fields. Featuring compact volumes of 50 to 125 pages, the series covers a range of content from professional to academic.

Typical publications can be:

- A timely report of state-of-the art methods
- An introduction to or a manual for the application of mathematical or computer techniques
- A bridge between new research results, as published in journal articles
- A snapshot of a hot or emerging topic
- An in-depth case study
- A presentation of core concepts that students must understand in order to make independent contributions

SpringerBriefs are characterized by fast, global electronic dissemination, standard publishing contracts, standardized manuscript preparation and formatting guidelines, and expedited production schedules.

On the one hand, **SpringerBriefs in Applied Sciences and Technology** are devoted to the publication of fundamentals and applications within the different classical engineering disciplines as well as in interdisciplinary fields that recently emerged between these areas. On the other hand, as the boundary separating fundamental research and applied technology is more and more dissolving, this series is particularly open to trans-disciplinary topics between fundamental science and engineering.

Indexed by EI-Compendex, SCOPUS and Springerlink.

Abdallah Hamed

Holographic Imaging Using Aperture Modulation

 Springer

Abdallah Hamed
Department of Physics
Faculty of Science
Ain Shams University
Cairo, Egypt

ISSN 2191-530X   ISSN 2191-5318 (electronic)
SpringerBriefs in Applied Sciences and Technology
ISBN 978-3-031-96988-1   ISBN 978-3-031-96989-8 (eBook)
https://doi.org/10.1007/978-3-031-96989-8

This Springer imprint is published by the registered company Springer Nature Switzerland AG
The registered company address is: Gewerbestrasse 11, 6330 Cham, Switzerland

If disposing of this product, please recycle the paper.

*I dedicate this book to the spirit of my
parents.*

Preface

Even since the invention of lasers in 1960, there has been a renaissance in the fields of optics and optical electronics. Optics and optical electronics are now being applied in all branches of science and engineering. The famous book *Introduction to Fourier Optics and Holography* by Goodman (1968) and *Laser Speckle and Applications in Optics* by Francon (1978), followed by my recent publications on the subject, led to the presentation of the two books published in the series *Springer Briefs in Applied Sciences and Technologies*. The content of the book, extracted from my recent publications, adds information on holographic imaging due to the modification that occurred in pupil distribution. This book is intended for graduate students in optical sciences.

The object of drafting a book on holographic imaging using pupils with linear and quadratic distributions is outlined below. When the circular aperture of uniform transmittance was replaced by modulated cracks, the point spread function (PSF) changed, and the cutoff spatial frequency was changed, leading to resolution improvement in the formed holographic image using a fixed diffuser.

This book highlights the formation of holographic images using modulated apertures comprised in eleven chapters. Chapter 1 summarizes some basic Fourier transformations used to compute the diffraction patterns of some famous objects. The basics of holographic photos are outlined in Chap. 2. Fourier holographic images obtained using argon plasma are investigated in Chap. 3. In addition, Fourier holographic images using modulated Hamming apertures are summarized in Chap. 4. The use of scanning holography with linear and quadratic apertures is discussed in Chaps. 5 and 6. The processing and segmentation of cancerous mammographic images using improved Fourier holograms are presented in Chap. 7, and Fourier holographic imaging of modulated apertures is presented in Chap. 8. The recognition of colored objects is presented in Chap. 9. Pattern recognition and processing of information were presented in Chap. 10. Finally, Chap. 11 is concerned with the operator algebra in complex coherent optical systems.

Cairo, EgyptAbdallah Hamed

Competing Interests The author has no competing interests to declare that are relevant to the content of this manuscript.

Contents

Chapter 1
Fourier Transform and Convolution Operations in Optical Diffraction

1.1 Introduction

Early in 1960, the impulse response or the point spread function (PSF) was computed for circular and annular apertures in many articles [1–4] using uniform illumination. The conventional microscope resolution is based on the resolution limit of the sole objective lens, while the resolution is improved for the confocal scanning laser microscope [5–8] since the two microscope objectives play equal roles in the resolution and the effective PSF is the multiplication of the PSF corresponding to each objective lens. Later, the resolution was further improved by investigating the PSF in the case of conical and linear [9, 10], quadratic and higher-order apertures [11]. In addition, twofold symmetry and fourfold symmetry apertures [12] are proposed, and the PSF is computed and compared with a circular aperture. Recently, different modulated apertures have been suggested, e.g., elliptical apertures, some deformed apertures [13], and graded index apertures [14], and the PSF has been computed in all cases. In addition, Hamming [15], linear quadratic [16], and longitudinal strips [17] have been investigated. The computations of the PSF corresponding to all the modulated apertures assume coherent uniform illumination.

In this chapter, we start with the Fourier transform properties. Then, the above studies of the modulated apertures for the computation of the PSF of a microscope objective are performed for non-uniform Gaussian illumination. PSF is computed from the Fourier transform of the multiplication of the Gaussian function and the aperture function. The result of the PSF computed from the convolution product of the PSF in the case of a uniform aperture with the Fourier spectrum of the Gaussian function. These results are compared with those obtained in the case of uniform illumination for all the considered modulated apertures. Finally, the results are discussed, and a conclusion is given.

© The Author(s), under exclusive license to Springer Nature Switzerland AG 2025

A. Hamed, *Holographic Imaging Using Aperture Modulation*,
SpringerBriefs in Applied Sciences and Technology,
https://doi.org/10.1007/978-3-031-96989-8_1

1.2   Basic Properties of the Fourier Transform

1. The definition of the Fourier transforms of a two-dimensional function g of two independent variables x and y is represented by

$$G(u, v) = \text{F.T.}\big[g(x, y)\big] = \iint\limits_{-\infty}^{\infty} g(x, y) \exp\big[-j2\pi(xu + yv)\big]dx\, dy. \qquad (1.1)$$

The transform is itself a complex-valued function of two independent variables, u and v, which are referred to as frequencies.

The inverse Fourier transform is similar, differing only in the sign of the exponent appearing in the integrand. It is written as follows:

$$g(x, y) = \text{F.T.}^{-1}[G(u, v)] = \iint\limits_{-\infty}^{\infty} G(u, v) \exp\big[j2\pi(xu + yv)\big]du\, dv. \qquad (1.2)$$

2. Linearity theorem:

$$\text{F.T.}\{\alpha f(x, y) + \beta g(x, y)\} = \alpha\text{F.T.}\{f(x, y)\} + \beta\text{F.T.}\{g(x, y)\}. \qquad (1.3)$$

That is, the transformation of a weighted sum of two or more functions is simply the weighted sum of their transformation.

3. Similarity theorem (scaling): If $\text{F.T.}\{g(x, y)\} = G(u, v)$, then:

$$\text{F.T.}\{g(ax, by)\} = \frac{1}{|ab|}G\Big(\frac{u}{a}, \frac{v}{b}\Big). \qquad (1.4)$$

That is, a stretch of the coordinates in the space domain (x, y) results in a contraction of the coordinates in the frequency domain (u, v), plus a change in the overall amplitude of the spectrum.

4. Shift theorem: If $\text{F.T.}\{g(x, y)\} = G(u, v)$, then

$$\text{F.T.}\{g(x - a, y - b)\} = G(u, v) \exp\big[-j2\pi(au + bv)\big]. \qquad (1.5)$$

That is, translation in the space domain introduces a linear phase shift in the frequency domain (u, v).

5. Rayleigh's theorem (Parseval's theorem):

$$\iint\limits_{-\infty}^{\infty} |g(x, y)|^2 dx\, dy = \iint\limits_{-\infty}^{\infty} |G(u, v)|^2 du\, dv. \qquad (1.6)$$

This leads to the conservation of the energy law since the integrand on the L.H.S. of this theorem can be interpreted as the energy contained in the waveform $g(x, y)$. A similar integrand in the R.H.S. represents the energy density in the frequency domain.

6. Convolution theorem:

If F.T.$\{g(x, y)\} = G(u, v)$ and F.T.$\{h(x, y)\} = H(u, v)$, then

$$\text{F.T}\left\{\iint_{-\infty}^{\infty} g(\xi, \eta)h(x - \xi, y - \eta)\mathrm{d}\xi \ \mathrm{d}\eta\right\} = \text{F.T}\{g(x, y) \otimes h(x, y)\}$$

$$= G(u, v) \cdot H(u, v). \qquad (1.7)$$

Hence, the Fourier transform of the convolution product in the space domain (x, y) results in a simple product of the Fourier spectrum corresponding to each function in the frequency domain (u, v).

7. Autocorrelation theorem:

$$\text{F.T}\left\{\iint_{-\infty}^{\infty} g(\xi, \eta)g^*(\xi - x, \eta - y)\mathrm{d}\xi \ \mathrm{d}\eta\right\} = |G(u, v)|^2. \qquad (1.8)$$

Similarly,

$$\text{F.T.}\{|g(x, y)|^2\} = \iint_{-\infty}^{\infty} G(\xi, \eta)G^*(\xi - u, \eta - v)\mathrm{d}\xi \ \mathrm{d}\eta \qquad (1.9)$$

This theorem may be regarded as a special case of the convolution theorem, in which we convolve $g(x, y)$ with $g^*(-x, -y)$.

8. Fourier integral theorem.

At each point of continuity of g,

$$\mathcal{F}\mathcal{F}^{-1}\{g(x, y)\} = \mathcal{F}^{-1}\mathcal{F}\{g(x, y)\} = g(x, y). \qquad (1.10)$$

At each point of discontinuity g, the two successive transforms yield the angular average of the values of g in a small neighborhood of that point. That is, the successive transformation and inverse transformation of a function yield that function again, except at points of discontinuity.

The above-mentioned theorems are of far more than just theoretical interest. They are used frequently since they provide the basic tools for the manipulation of Fourier transformers and can save enormous amounts of work in the solution of Fourier analysis problems.

1.3 Fourier Transform Examples Are Used in Optics and Image Processing

In the following subsections, we calculate the Fourier transform corresponding to the rectangular aperture, Gaussian function, two displaced pinholes (Young's experiment), a random array of identical tiny objects, and circular and annular apertures.

1.3.1 Fourier Transforms of the Rectangular Aperture

First, consider a rectangular aperture with an amplitude transmittance given by

$$t_A(\xi, \eta) = \text{rect}\left(\frac{\xi}{2w_x}\right)\text{rect}\left(\frac{\eta}{2w_y}\right). \tag{1.11}$$

The constants w_x and w_y are the half-widths of the aperture in the ξ and η directions, respectively. If the aperture is illuminated by a unit amplitude, normally incident, monochromatic plane wave, then the field distribution across the aperture is equal to the transmittance function t_A. Thus, the Fraunhofer diffraction pattern is [2]

$$U(x, y) = \frac{e^{jkz}}{j\lambda z}e^{\frac{jk}{2z}(x^2+y^2)}\text{F.T.}\{U(\xi, \eta)\}. \tag{1.12}$$

Noting that

$$\text{F.T.}\{U(\xi, \eta)\} = A\, sinc(2w_x f_x)\, sinc(2w_y f_y) \tag{1.13}$$

where A is the area of the aperture ($A = 4w_x w_y$), and the spatial frequencies are as follows:

$$f_x = \frac{x}{\lambda f} \text{ and } f_y = \frac{y}{\lambda f}$$

$$U(x, y) = \frac{e^{jkz}}{j\lambda z}e^{\frac{jk}{2z}(x^2+y^2)}A\, sinc(2w_x x/\lambda f)\, sinc(2w_y y/\lambda f) \tag{1.14}$$

and

$$I(x, y) = \frac{A^2}{\lambda^2 f^2}\, sinc^2(2w_x f_x)\, sinc^2(2w_y f_y). \tag{1.15}$$

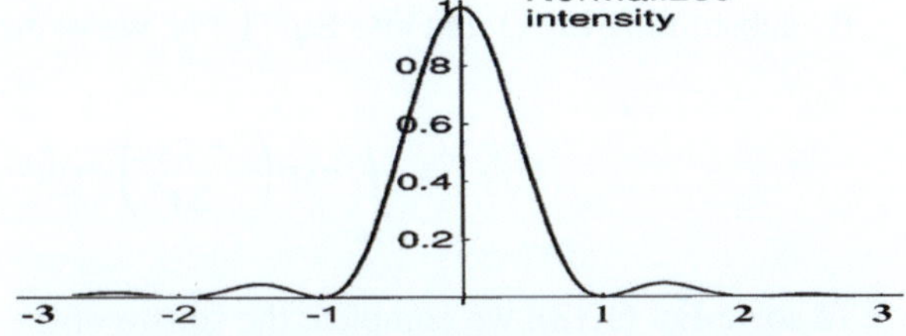

Fig. 1.1 Cross-section of the Fraunhofer diffraction pattern of a rectangular aperture

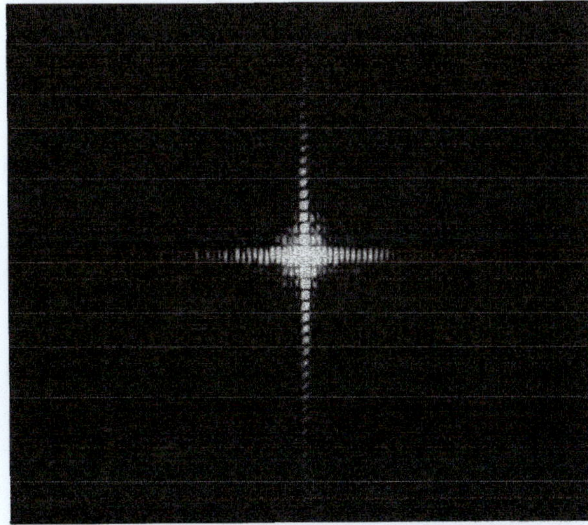

Fig. 1.2 Fraunhofer diffraction pattern of a rectangular aperture, $\frac{w_x}{w_y} = 2$

Figure 1.1 shows a cross-section of the Fraunhofer intensity pattern along the x-axis. Note that the width of the main lobe (i.e., the distance between the first two zeros) is $\Delta x = \lambda z / w_x$ (Fig. 1.2).

1.3.2   Fourier Transform of the Gaussian Function

We assume a Gaussian function in the time domain represented as follows:

$$g(t) = \exp\left(-\frac{t^2}{2\sigma^2}\right). \tag{1.16}$$

The Fourier transform of Eq. (1.16) is written as follows:

$$G(\omega) = \int_{-\infty}^{\infty} g(t) \exp(-j\omega t)\,\mathrm{d}t. \tag{1.17}$$

By substituting Eq. (1.16) into Eq. (1.17), we write:

$$G(\omega) = \int_{-\infty}^{\infty} \exp\left(-\frac{t^2}{2\sigma^2}\right) \exp(-j\omega t)\mathrm{d}t. \qquad (1.18)$$

To solve Eq. (1.18), we complete the square of the exponent inside the integral; then, we write:

$$G(\omega) = \exp\left(-\frac{\omega^2\sigma^2}{2}\right) \int_{-\infty}^{\infty} \exp\left\{-\left(\frac{t}{\sigma\sqrt{2}} + \frac{j\omega\sigma}{\sqrt{2}}\right)^2\right\}\mathrm{d}t. \qquad (1.19)$$

Letting $\frac{t}{\sigma\sqrt{2}} + \frac{j\omega\sigma}{\sqrt{2}} = t'$, $\mathrm{d}t = \sigma\sqrt{2}\mathrm{d}t'$ and substituting into Eq. (1.19), we obtain:

$$G(\omega) = \sigma\sqrt{2}\exp\left(-\frac{\omega^2\sigma^2}{2}\right) \int_{-\infty}^{\infty} \exp\left[-(t')^2\right]\mathrm{d}t'. \qquad (1.20)$$

The Gaussian integral, known as the Euler-Poisson integral, is:

$$\int_{-\infty}^{\infty} \exp\left[-(t')^2\right]\mathrm{d}t' = \sqrt{\pi}.$$

Substitute in Eq. (1.20), we obtain another Gaussian function for the Fourier transform of the original Gaussian function represented symbolically as follows:

$$G(\omega) = \sigma\sqrt{2\pi}\exp\left(-\frac{\omega^2\sigma^2}{2}\right) = \text{F.T.}\left\{\exp\left(-\frac{t^2}{2\sigma^2}\right)\right\}. \qquad (1.21)$$

1.3.3   *Fourier Transform of Two Displaced Pinholes (Young's Experiment)*

Two symmetrical pinholes are assumed to be placed in the plane (x, y), where the pinholes are aligned along the x-coordinate at distances $x = \pm d/2$.

In this case, the pinholes that represent the object information are represented as follows:

$$g(x, y) = \delta\left(x - \frac{d}{2}, y\right) + \delta\left(x + \frac{d}{2}, y\right). \qquad (1.22)$$

The Fourier transform of Eq. (1.22) is in the observation plane (u, v) at a great distance z compared with the distance (d) between the pinholes $(z \gg d)$.

This condition is fulfilled to record the Fraunhofer diffraction (Fourier plane) or use a converging lens and record in its focal plane.

Hence, operating the transformation of Eq. (1.22), we obtain:

$$\overline{g}(u, v) = \int\limits_{-\infty}^{\infty} \int\limits_{-\infty}^{\infty} \left[\delta\left(x - \frac{d}{2}, y\right) + \delta\left(x + \frac{d}{2}, y\right) \right] \exp\left[-\frac{j2\pi}{\lambda f}(xu + yv) \right] dx\, dy.$$

$$(1.23)$$

It is decomposed into two integrals as follows:

$$\overline{g}(u, v) = I_1 + I_2$$

$$I_1 = \int\limits_{-\infty}^{\infty} \int\limits_{-\infty}^{\infty} \delta\left(x - \frac{d}{2}, y\right) \exp\left[-\frac{j2\pi}{\lambda f}(xu + yv) \right] dx\, dy \qquad (1.24)$$

$$I_2 = \int\limits_{-\infty}^{\infty} \int\limits_{-\infty}^{\infty} \delta\left(x + \frac{d}{2}, y\right) \exp\left[-\frac{j2\pi}{\lambda f}(xu + yv) \right] dx\, dy. \qquad (1.25)$$

The 1st integral represented by Eq. (1.24) is solved as follows:

$$I_1 = \exp\left(-\frac{j\pi d}{\lambda f}\right) \int\limits_{-\infty}^{\infty} \int\limits_{-\infty}^{\infty} \delta\left(x - \frac{d}{2}, y\right) \exp\left\{ -\frac{j2\pi}{\lambda f}\left[\left(x - \frac{d}{2}\right)u + yv \right] \right\} dx\, dy.$$

$$(1.26)$$

Let $x - d/2 = x'$, and $dx = dx'$, and substituting in the above equation, we obtain:

$$I_1 = \exp\left(-\frac{j\pi d}{\lambda f}u\right) \int\limits_{-\infty}^{\infty} \int\limits_{-\infty}^{\infty} \delta(x', y) \exp\left[-\frac{j2\pi}{\lambda f}(x'u + yv) \right] dx\, dy. \qquad (1.27)$$

Since the Fourier transform of the Dirac-delta function is unity, the above integral gives the following result:

$$I_1 = \exp\left(-\frac{j\pi d}{\lambda f}u\right). \qquad (1.28)$$

Similarly,

$$I_2 = \exp\left(+\frac{j\pi d}{\lambda f}u\right). \qquad (1.29)$$

By substituting Eqs. (1.28) and (1.29) into Eq. (1.23), we obtain:

$$
\begin{aligned}
\overline{g}(u, v) &= \exp\left(-\frac{j\pi d}{\lambda f}u\right) + \exp\left(+\frac{j\pi d}{\lambda f}u\right) \\
&= 2\cos\left(\frac{\pi d}{\lambda f}u\right).
\end{aligned}
\tag{1.30}
$$

The corresponding intensity is the familiar $\cos^2$ function valid for two beam interferences represented as follows:

$$
I(u, v) = 4\cos^2\left(\frac{\pi d}{\lambda f}u\right).
\tag{1.31}
$$

1.3.4   Fourier Transform of a Random Array of Identical Tiny Objects

A random array of two-dimensional delta functions is written as follows:

$$
\text{Rand}(x, y) = \sum_{i=1}^{N} \delta(x - x_i, y - y_i)
\tag{1.32}
$$

where x and y are elements in the square matrix of dimensions $N \times N$.

The Fourier transform is easily computed from Eq. (1.32) as follows:

$$
\overline{g}(u, v) = \int_{-\infty}^{\infty} \int_{-\infty}^{\infty} \sum_{i=1}^{N} \delta(x - x_i, y - y_i) \exp\left[-\frac{j2\pi}{\lambda f}(xu + yv)\right] dx \, dy.
\tag{1.33}
$$

By solving the transformation in Eq. (1.33), we obtain:

$$
\overline{g}(u, v) = \sum_{i=1}^{N} \exp\left[-\frac{j2\pi}{\lambda f}(x_i u + y_i v)\right].
\tag{1.34}
$$

The equation represents randomly varying sinusoids that appear as noisy patterns located in the (u, v) plane.

1.3.5 Fourier Transforms of the Circular Aperture

The circular aperture of the transmission function is defined as follows:

$$P(x, y) = 1 \text{ for } \left| \frac{r}{r_0} \right| \leq 1$$

$$= 0 \text{ for } \left| \frac{r}{r_0} \right| \geq 1. \tag{1.35}$$

We solve the problem in two dimensions, using polar coordinates, as follows.

Returning now to formula (1.35), and applying the two-dimensional Bessel-Fourier transform, leads us to write the point spread function as follows:

$$h_1(\rho) = \int_0^{2\pi} \int_0^1 \exp\left(-\frac{j2\pi}{\lambda f}\rho r \cos\theta\right) r \, dr \, d\theta = 2\pi \int_0^1 r \, J_0\left(\frac{k\rho r}{f}\right) dr \tag{1.36}$$

with $k = 2\pi/\lambda$, is the propagation constant.

We rewrite Eq. (1.36), for the objective lens, as follows:

$$h_1(w) = 2\pi \left(\frac{f}{k\rho}\right)^2 \int_0^W w \, J_0(w) dw \tag{1.37}$$

with $W = \frac{k r_0 \rho}{f}$, and $w = k\rho r/f$.

We solve the integral in Eq. (1.37) using the following recurrence formula:

$$w\frac{d}{dw}[J_n(w)] + J_n(w) = w J_{n-1}(w). \tag{1.38}$$

We integrate Eq. (1.38) with $n = 1$ to obtain:

$$w\frac{d}{dw}[J_1(w)] + J_1(w) = w J_0(w)$$

$$\rightarrow \frac{d}{dw}[w J_1(w)] = w J_0(w). \tag{1.39}$$

Hence, we obtain, by integrating Eq. (1.39), the following result:

$$w J_1(w) = \int_0^W w J_0(w) dw = W J_1(W). \tag{1.40}$$

By substituting Eq. (1.40) into Eq. (1.37), we obtain:

$$h_1(w) = 2\pi \left(\frac{f}{k\rho}\right)^2 \cdot WJ_1(W)$$

$$= \pi r_0^2 \left[2\frac{J_1(W)}{W}\right]. \tag{1.41}$$

In the center of the diffraction pattern with $\rho = 0$, the properties of the Bessel function give

$$\lim_{w \to 0}\left[2\frac{J_1(W)}{W}\right] = 1. \tag{1.42}$$

By substituting Eq. (1.42) into Eq. (1.41), we obtain the area of the circular aperture πr_0^2.

1.3.6   *Fourier Transforms of the Annular Aperture*

The Fourier transform of the annular aperture, using Eq. (1.42), is computed as follows:

$$h_{\text{annul.}}(r) = 2\pi r_0^2 \left[\frac{J_1(W)}{W} - \epsilon^2 \frac{J_1(\epsilon W)}{\epsilon W}\right]; \; \epsilon = \frac{b}{a}, \tag{1.43}$$

where a: is the external radius and b: is the internal radius of the circular aperture.

1.4   Diffraction of Sinusoidal Amplitude Grating (Application of Fourier Transform and Convolution Operations)

The transmission functions in the above examples of rectangular and circular apertures have the following forms:

$$T(\xi, \eta) = 1, \quad \text{in the aperture}$$
$$= 0, \quad \text{outside the aperture.}$$

In this example, we compute the diffraction originating from an amplitude grating using Fourier transform and convolution operations.

Hence, the diffracting screen has a thin sinusoidal amplitude grating of transmittance function represented as follows:

Fig. 1.3 Amplitude transmittance function of the sinusoidal amplitude grating

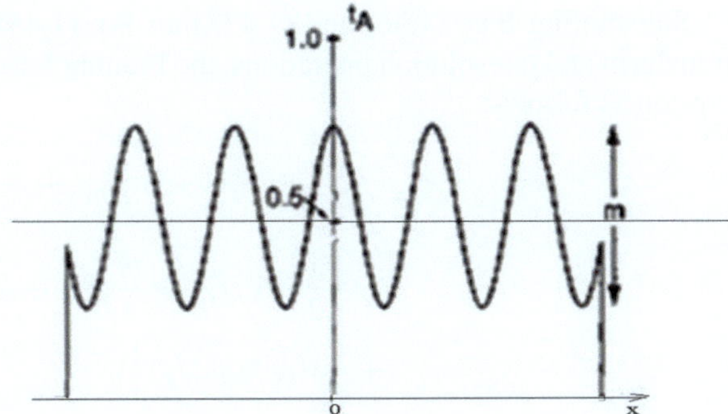

$$T_A(\xi, \eta) = \left[\frac{1}{2} + \frac{m}{2}\cos(2\pi f_0\xi)\right]\text{rect}\left(\frac{\xi}{2w}\right)\text{rect}\left(\frac{\eta}{2w}\right) \tag{1.44}$$

where we assume that the grating structure is bounded by a square aperture of width $2w$. The parameter m represents the peak-to-peak change in the amplitude transmittance across the screen, and f_0 is the spatial frequency of the grating. Figure 1.3 shows a cross-section of the grating amplitude transmittance function.

If the screen is normally illuminated by a unit amplitude plane wave originating from a spatially filtered laser beam, the field distribution across the aperture is equal to $T(\xi, \eta)$. To determine the Fraunhofer diffraction pattern of Eq. (1.44), we apply the following formula:

$$h(x, y) = \frac{e^{jkz}}{j\lambda z} e^{\frac{jk}{2z}(x^2+y^2)} \int\!\!\!\int\limits_{-\infty}^{\infty} T_A(\xi, \eta) \times \exp\left\{-\frac{j2\pi}{\lambda z}(x\xi + y\eta)\right\} d\xi\, d\eta. \tag{1.45}$$

Hence, we apply the Fourier transform Eqs. (1.45)–(1.44) and neglect the exponential complex term before the integration since quadratic detection in the Fourier plane is equal to unity; hence, it does not affect the detected image.

We compute the F.T. corresponding to each function in Eq. (1.44) separately as follows:

$$\text{F.T.}\left[\frac{1}{2} + \frac{m}{2}\cos(2\pi f_0\xi)\right] = \frac{1}{2}\delta(f_X, f_Y) + \frac{m}{4}\delta(f_X + f_0, f_Y)$$
$$+ \frac{m}{4}\delta(f_X + f_0, f_Y) \tag{1.46}$$

$$\text{F.T.}\left[\text{rect}\left(\frac{\xi}{2w}\right)\text{rect}\left(\frac{\eta}{2w}\right)\right] = A\, sinc(2wf_X)\, sinc(2wf_Y). \tag{1.47}$$

where $A = 4w^2$ signifies the area of the aperture bounding the grating.

Substituting Eqs. (1.46) and (1.47) into Eq. (1.45), and making use of Fourier transform and convolution operations, the Fraunhofer diffraction pattern can now be written as follows:

$$
\begin{aligned}
h(u, v) = {} & \frac{A}{j2\lambda z} e^{jkz} e^{\frac{jk}{2z}(x^2+y^2)} \; sinc(2wf_X) \; sinc(2wf_Y) \\
& \otimes \left[\frac{1}{2}\delta(f_X, f_Y) + \frac{m}{4}\delta(f_X + f_0, f_Y) \right. \\
& \left. + \frac{m}{4}\delta(f_X + f_0, f_Y) \right].
\end{aligned}
\tag{1.48}
$$

The convolution theorem can be used to write Eq. (1.48) as follows:

$$
\begin{aligned}
h(x, y) = {} & \frac{A}{j2\lambda z} e^{jkz} e^{\frac{jk}{2z}(x^2+y^2)} \; sinc(2wf_Y) \{ \; sinc(2wf_X) \\
& + \frac{m}{2} \; sinc\left[2w(f_X + f_0)\right] \\
& + \frac{m}{2} \; sinc\left[2w(f_X - f_0)\right] \}.
\end{aligned}
\tag{1.49}
$$

The PSF or Fraunhofer diffraction can be written as

$$
\begin{aligned}
h(x, y) = {} & \frac{A}{j2\lambda z} e^{jkz} e^{\frac{jk}{2z}(x^2+y^2)} \; sinc\left(\frac{2wy}{\lambda z}\right) \left\{ \; sinc\left(\frac{2wx}{\lambda z}\right) \right. \\
& + \frac{m}{2} \; sinc\left[\frac{2w}{\lambda z}(x + f_0\lambda z)\right] \\
& \left. + \frac{m}{2} \; sinc\left[\frac{2w}{\lambda z}(x - f_0\lambda z)\right] \right\}.
\end{aligned}
\tag{1.50}
$$

Finally, the corresponding intensity distribution is found by taking the squared magnitude of Eq. (1.50). Note that if there are many grating periods within the aperture $f \gg 1/w$, there will be negligible overlap of the three sinc functions, allowing the intensity to be calculated as the sum of the squared magnitudes of the three terms in Eq. (1.50). The intensity is then given by:

$$
\begin{aligned}
I(x, y) \approx {} & \left[\frac{A}{2\lambda z}\right]^2 sinc^2\left(\frac{2wy}{\lambda z}\right) \left\{ \; sinc^2\left(\frac{2wx}{\lambda z}\right) \right. \\
& + \frac{m^2}{4} \; sinc^2\left[\frac{2w}{\lambda z}(x + f_0\lambda z)\right] \\
& \left. + \frac{m^2}{4} \; sinc^2\left[\frac{2w}{\lambda z}(x - f_0\lambda z)\right] \right\}.
\end{aligned}
\tag{1.51}
$$

The intensity pattern is shown in Fig. 1.4. Note that some of the incident light is absorbed by the grating, and in addition, the sinusoidal transmittance variation

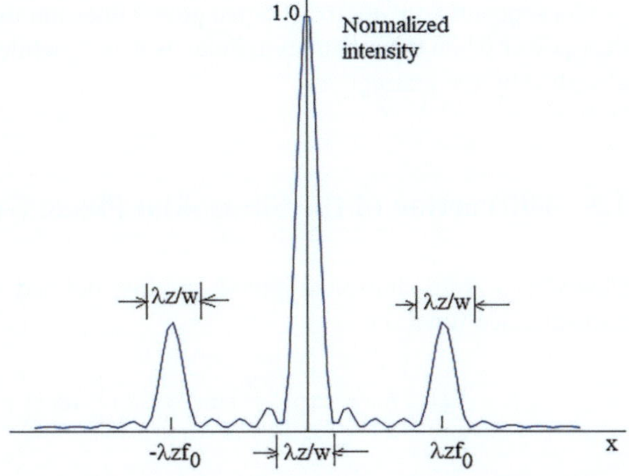

Fig. 1.4 The normalized diffracted intensity vs. the x-axis is shown in the plot using Eq. (1.51). It corresponds to the Fraunhofer diffraction from a thin sinusoidal amplitude grating

across the aperture has deflected some of the energy out of the central diffraction pattern into two additional side patterns. The central diffraction pattern is called the zero-order Fraunhofer pattern, while the two side patterns are called the first orders. The spatial separation of the first orders from the zero order is $f_0\,\lambda z$, while the width of the main lobe of all orders is $\lambda z/w$.

Another important practical interest in both holography and optical information processing is the diffraction efficiency of the grating. It is defined as the fraction of the incident optical power that appears in a single diffraction order (usually the $+1$ order) of the grating.

The diffraction efficiency for the grating of interest can be deduced from Eq. (1.48). The fraction of power appearing in each diffraction order can be found by squaring the coefficients of the delta functions in this representation. It is the delta function that determines the power in each order, not the sinc functions that simply spread these impulses. From Eq. (1.48), we can conclude that the diffraction efficiencies η_0, and η_{+1}, η_{-1} associated with the three diffraction orders are given by:

$$\eta_0 = 0.25, \eta_{+1} = \frac{m^2}{16}, \eta_{-1} = \frac{m^2}{16}. \tag{1.52}$$

Thus, a single first diffraction order carries at most $1/16 = 6.25\%$ of the incident power, a rather small fraction. If the frequencies of the three orders are summed, only $1/4 + \frac{m^2}{8}$ of the total is accounted for. The rest is lost through absorption by the grating.

The total diffracted power is the sum of the three diffracted orders:

$$P_T = \frac{1}{4} + \frac{m^2}{8} = 37.5\%, \quad \text{for } m = 1. \tag{1.53}$$

Consequently, the total diffracted power does not exceed 37.5% of the total incident power when the modulation index is $m = 1$, while a major part of the energy is absorbed by the grating.

1.5   Diffraction of the Sinusoidal Phase Grating

Consider a thin sinusoidal phase grating defined by the complex amplitude transmittance function:

$$T_A(\xi, \eta) = \exp\left[\frac{jm}{2}\sin(2\pi f_0 \xi)\right]\,\mathrm{rect}\left(\frac{\xi}{2w}\right)\mathrm{rect}\left(\frac{\eta}{2w}\right). \qquad (1.54)$$

The parameter m represents the peak-to-peak excursion of the phase delay.

If the screen is normally illuminated by a unit amplitude plane wave originating from a spatially filtered laser beam, the field distribution behind the aperture is given precisely by Eq. (1.54).

The analysis is simplified using the identity:

$$\exp\left[\frac{jm}{2}\sin(2\pi f_0 \xi)\right] = \sum_{q=-\infty}^{\infty} J_q\left(\frac{m}{2}\right)\exp(j2\pi q f_0 \xi), \qquad (1.55)$$

where J is a Bessel function of the first kind of order q.

Thus, the diffraction from the phase grating is computed from Eqs. (1.54) and (1.55) using the definition of F.T. as follows:

$$\mathrm{F.T.}\left\{\exp\left[\frac{jm}{2}\sin(2\pi f_0 \xi)\right]\right\} = \sum_{q=-\infty}^{\infty} J_q\left(\frac{m}{2}\right)\mathrm{F.T.}\{\exp(j2\pi q f_0 \xi)\}$$

$$= \sum_{q=-\infty}^{\infty} J_q\left(\frac{m}{2}\right)\delta(f_X - q f_0, f_Y) \qquad (1.56)$$

and

$$U(f_X, f_Y) = \mathrm{F.T.}\{T_A(\xi, \eta)\}$$

$$= \left[A\,\mathrm{sinc}(2w f_X)\,\mathrm{sinc}(2w f_Y)\right]$$

$$\otimes \sum_{q=-\infty}^{\infty} J_q\left(\frac{m}{2}\right)\delta(f_X - q f_0, f_Y)$$

$$= \sum_{-\infty}^{\infty} A J_q\left(\frac{m}{2}\right)\mathrm{sinc}\left[2w(f_X - q f_0)\right]\,\mathrm{sinc}(2w f_Y). \qquad (1.57)$$

Thus, the field strength in the Fraunhofer diffraction pattern becomes:

$$U(x, y) = \frac{A}{j\lambda z} e^{jkz} e^{\frac{jk}{2z}(x^2+y^2)}$$

$$\times \sum_{-\infty}^{\infty} AJ_q\left(\frac{m}{2}\right) sinc\left[\frac{2w}{\lambda z}(x - qf_0\lambda z)\right] sinc\left(2w\frac{y}{\lambda z}\right). \qquad (1.58)$$

If we again assume that there are many grating periods within the bounding aperture ($f \gg 1/w$), there is a negligible overlap of the various diffracted terms, and the corresponding intensity becomes:

$$I(x, y) \approx \left(\frac{A}{\lambda z}\right)^2 \sum_{-\infty}^{\infty} J_q^2\left(\frac{m}{2}\right) sinc^2\left[\frac{2w}{\lambda z}(x - qf_0\lambda z)\right] sinc^2\left(\frac{2wy}{\lambda z}\right). \qquad (1.59)$$

Thus, the introduction of the sinusoidal phase grating deflected the energy out of the zero order into a multitude of higher orders. The peak intensity of the q order is:

$$I_{max} = \left[A\frac{J_q\left(\frac{m}{2}\right)}{\lambda z}\right]^2. \qquad (1.60)$$

The displacement of that order from the center of the diffraction pattern is $qf_0\lambda z$. Figure 1.5 shows a cross-section of the intensity pattern when the phase delay is $m = 8$ radians. Note that the strengths of the diffracted orders are symmetric about the zero order.

The diffraction efficiency of the thin phase grating is obtained by squaring the coefficients in Eq. (1.56). Thus, the diffraction efficiency of the q order corresponding to this grating becomes:

$$\eta_q = J_q^2\left(\frac{m}{2}\right). \qquad (1.61)$$

Fig. 1.5 The normalized diffracted intensity vs. the x- axis for a thin sinusoidal phase grating. The ±1 order nearly vanishes in this example where $m = 8$

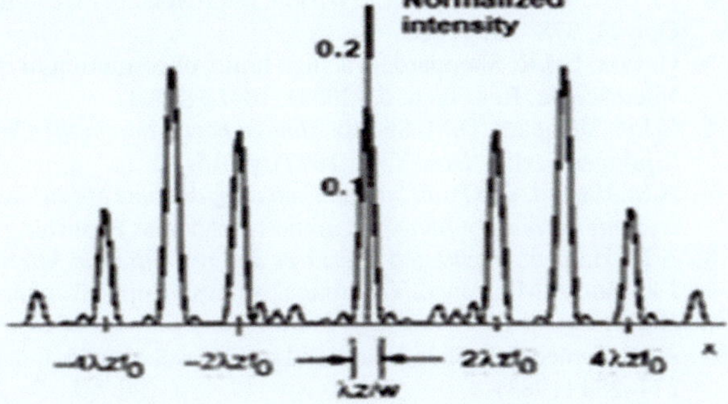

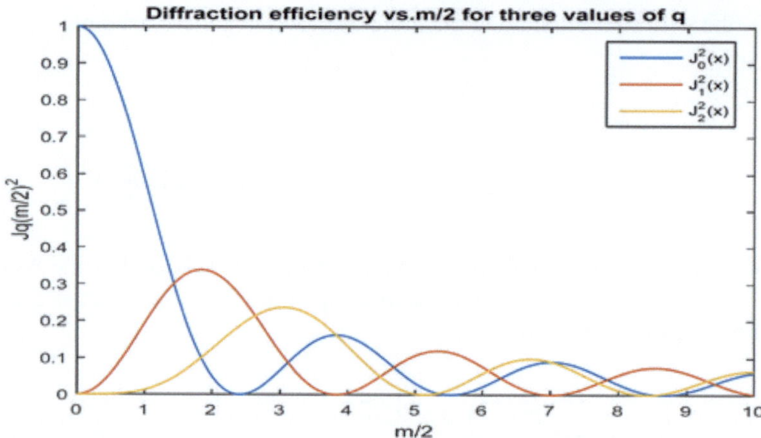

Fig. 1.6 Diffraction efficiency versus the modulation index $m/2$ for three different orders $q = 0, 1$, and 2 Using Eq. (1.61)

The diffraction efficiency given by Eq. (1.61) is plotted in Fig. 1.6 for three different orders $q = 0$, 1, and 2.

The largest possible diffraction efficiency is found in the first order $q = 1, -1$, computed from the maximum value of J_1^2, which equals nearly 33.8% compared with 6.25% for the same diffracted order. In addition, no power is absorbed by this grating, and therefore, the sum of the powers appearing in all orders remains constant and equal to the incident power. In this example, the two diffracted orders corresponding to $q = 1$ and $q = 2$ only give a total diffraction efficiency of nearly 59% (Fig. 1.6).

References

1. M. Minsky, Memoir on inventing the confocal microscope. Scanning **10**, 128–138 (1988)
2. J.W. Goodman, *Introduction to Fourier Optics* (McGraw Hill Series, 1996)
3. C.J.R. Sheppard, A. Choudhary, Opt. Acta **24**, 1051–1059 (1977)
4. I.J. Cox, C.J.R. Sheppard, T. Wilson, Improvement in resolution by confocal microscopy. Appl. Opt. **21**, 778–781 (1982)
5. G. Cox, C.J.R. Sheppard, Practical limits of resolution in confocal and nonlinear microscopy. Microscopic. Res. Tech. **63**(2004), 18–22 (2004)
6. C.J.R. Sheppard, D.M. Shotton, *Image Formation in the Confocal Laser Scanning Microscope* (Springer-Verlag, New York, 1997), pp.15–31
7. A.M. Hamed, *The Point Spread Function for Some Modulated Apertures Application on Speckle and Interferometry Images* (Lambert Academic Publishing (LAP), 2017)
8. A.M. Hamed, *Modulated apertures and resolution in Microscopy* (Springer, 2023)
9. J.J. Clair, A.M. Hamed, Theoretical studies on optical coherent microscope. Optik **64**, 133–141 (1983)
10. A.M. Hamed, J.J. Clair, Image and super-resolution in optical coherent microscopes. Optik **64**, 277–284 (1983)
11. A.M. Hamed, J.J. Clair, Studies on optical properties of confocal scanning optical microscope using pupils with radially transmission distribution. Optik **65**, 209–218 (1983)

12. A.M. Hamed, Resolution, and contrast in confocal optical scanning microscope. Opt. Laser Technol. **16**, 93–96 (1984)
13. A. M. Hamed, Discrimination between speckle images using diffusers modulated by some deformed apertures: simulations. Opt. Eng. **50**, 1–7 (2011) https://doi.org/10.1117/1.3530085
14. A.M. Hamed, Study of graded index and truncated apertures using speckle images. Precis. Instr. Mech. **3**, 144–152 (2014)
15. A.M. Hamed, T.A. Al-Saeed, Image analysis of modified Hamming aperture: application on confocal microscopy and holography. J. Mod. Opt. **62**, 801–810 (2015). https://doi.org/10.1080/09500340.2015.1007102
16. A.M. Hamed, Improvement of point spread function (PSF) using linear-quadratic aperture. Optik **131**, 838–849 (2017). https://doi.org/10.1016/j.ijleo.2016.11.201
17. A.M. Hamed et al., The point spread function using longitudinal black and white strips inside a circular aperture. Int. J. Photon. Opt. Technol. **3**, 1–9 (2017)

Chapter 2
Basics of Holography

2.1 Introduction

For a better understanding of the process, it is necessary to understand interference and diffraction. Interference occurs when one or more wavefronts are superimposed. Diffraction occurs when a wavefront encounters an object. The process of producing a holographic reconstruction is explained below purely in terms of interference and diffraction. This approach is somewhat simplified but accurate enough to provide an understanding of how the holographic process works.

2.2 Complex Objects

To record a hologram of a complex object, a laser beam is first split into two beams of light. One beam illuminates the object, which then scatters light onto the recording medium. According to diffraction theory, each point in the object acts as a point source of light, so the recording medium can be considered to be illuminated by a set of point sources located at varying distances from the medium.

The second (reference) beam illuminates the recording medium directly. Each point source wave interferes with the reference beam, giving rise to its sinusoidal zone plate in the recording medium. The resulting pattern is the sum of all these "zone plates", which are combined to produce a random (speckle) pattern.

When the hologram is illuminated by the original reference beam, each of the individual zone plates reconstructs the object wave that produced it, and these individual wavefronts are combined to reconstruct the whole of the object beam. The viewer perceives a wavefront that is identical to the wavefront scattered from the object onto the recording medium so that it appears that the object is still in place even if it has been removed.

© The Author(s), under exclusive license to Springer Nature Switzerland AG 2025

A. Hamed, *Holographic Imaging Using Aperture Modulation*,

SpringerBriefs in Applied Sciences and Technology,

https://doi.org/10.1007/978-3-031-96989-8_2

2.3 Mathematical Model

A single-frequency light wave can be modeled by a complex number, U, which represents the electric or magnetic field of the light wave. The amplitude and phase of the light are represented by the absolute value and angle of the complex number. The object and reference waves at any point in the holographic system are given by U_O and U_R. The combined beam is given by $U_0 + U_R$. The energy of the combined beams is proportional to the square of the magnitude of the combined waves as

$$I(x, y) = |U_0 + U_R|^2. \tag{2.1}$$

The complex amplitude diffracted from the object over the holographic plane is represented as follows:

$$U_0(x, y) = a_0(x, y) \exp\big[j\theta(x, y)\big]. \tag{2.2}$$

The complex amplitude of the reference or carrier wave incident upon the holographic plate is written as follows:

$$U_R(x, y) = a_R(x, y) \exp\big[j\Psi(x, y)\big] \tag{2.3}$$

If a photographic plate is exposed to the two beams and then developed, its transmittance, T, is proportional to the light energy that was incident on the plate and is given by:

$$T(x, y) = k \cdot I(x, y) \tag{2.4}$$

where k is a constant.

Alternatively, after processing the photographic plate under linear conditions, the transmitted amplitude becomes proportional to the recorded intensity, i.e.,

$$T(x, y) = \big[I(x, y)\big]^{-\gamma/2}. \tag{2.5}$$

γ is the slope of the characteristic curve of photographic emulsion.

When the developed plate is illuminated by the reference beam, the light transmitted through the plate, U_H, is equal to the transmittance, T, multiplied by the reference beam amplitude, U_R, giving Eq. (2.6).

$$U_H(x, y) = U_R \cdot T = kU_R I(x, y). \tag{2.6}$$

By substituting Eq. (2.1) into Eq. (2.6), we obtain the following result:

$$U_H(x, y) = kU_R\big\{|U_0|^2 + |U_R|^2 + U_R U^*(\text{object}) + U_R U(\text{object})\big\}. \tag{2.7}$$

By substituting Eq. (2.5) into Eq. (2.6), we obtain the following result:

$$U_H(x, y) = U_R T(x, y) = U_R\left[I(x, y)\right]^{-\gamma/2}.$$ (2.8)

By substituting Eqs. (2.2, 2.3) into Eq. (2.8), we obtain the following:

$$U_H(x, y) = U_R\left\{|U_0|^2 + |U_R|^2 + U_R U^*(\text{object}) + U_R U(\text{object})\right\}^{-\gamma/2}$$

$$U_H(x, y) = |U_R|^{-\gamma}\left\{1 + |U_0/U_R|^2 + U_R U^*(\text{object})/|U_R|^2\right.$$

$$\left. + U_R U(\text{object})/|U_R|^2\right\}^{-\gamma/2}.$$ (2.9)

Assuming the irradiance of the reference beam $|U_R|^2 = |a_R|^2 = 1$, and taking into consideration that $\frac{\gamma}{2}\left|\frac{U_0}{U_R}\right|^2$ is negligible for $U_0/U_R \ll 1$, Eq. (2.9) can be approximated as:

$$U_H(x, y) = |U_R|^{-\gamma}\left\{1 - \frac{\gamma}{2}\left|\frac{U_0}{U_R}\right|^2 - \frac{\gamma}{2}U_R\frac{U^*(\text{object})}{|U_R|^2} - \frac{\gamma}{2}U_R^*\frac{U(\text{object})}{|U_R|^2}\right\}.$$ (2.10)

We finally obtained the following result:

$$U_H(x, y) = |U_R|^{-\gamma}\left\{1 - \frac{\gamma}{2}U_R\frac{U^*(\text{object})}{|U_R|^2} - \frac{\gamma}{2}U_R^*\frac{U(\text{object})}{|U_R|^2}\right\}.$$ (2.11)

Consequently, we obtain an expression containing both the amplitude and phase of the diffracted wave from the object. By substituting Eqs. (2.2, 2.3) into Eq. (2.11), we can rewrite the above expression with the appearance of the interference aspect as follows:

$$U_H(x, y) = \left\{1 - \gamma a_0(x, y)\cos\left[\theta(x, y) - \Psi\right]\right\}.$$ (2.12)

The processed hologram is illuminated by a reconstruction laser beam such as the reference beam used during the recording process. The transmitted amplitude becomes:

$$U_t = U_{\text{reconst}}\left\{1 - \frac{\gamma}{2}U(\text{reference}) \cdot U^*(\text{object})\right.$$

$$\left. - \frac{\gamma}{2}U^*(\text{reference}) \cdot U(\text{object})\right\}$$ (2.13)

where $|U_R|^2 = 1$.

$$U_t = U_{\text{reconst}} - \frac{\gamma}{2}U_0 U_R^* U_{\text{reconst}} - \frac{\gamma}{2}U_0^* U_R U_{\text{reconst}}$$ (2.14)

$$U_t = U_{\text{reconst}} - \frac{\gamma}{2}U_0 - \frac{\gamma}{2}U_0^* \cdot (U_R)^2. \tag{2.15}$$

Consequently, the hologram reconstructs the following three waves:

1. The first term U_{reconst} corresponds to the direct transmitted wave. The wave is perturbed by the term $-\frac{\gamma}{2}\left|\frac{U_0}{U_R}\right|^2$, which was neglected.
2. Except for the factor $-\frac{\gamma}{2}$, the second term represents the direct image corresponding to the diffracted wave from the object.
3. The third term $-\frac{\gamma}{2}U_0^* \cdot (U_R)^2$ represents the conjugate image.

U_H has four terms, each representing a light beam emerging from the hologram. The first of these is proportional to $|U_0|^2$. This is the reconstructed object beam, which enables a viewer to "see" the original object even when it is no longer present in the field of view. The second and third beams are modified versions of the reference beam. The fourth term is the "conjugate object beam". It has the reverse curvature to the object beam itself and forms a real image of the object in space beyond the holographic plate. When reference and object beams are incident on the holographic recording medium at significantly different angles, the virtual, real, and reference wavefronts all emerge at different angles, enabling the reconstructed object to be observed.

2.4 Recording a Hologram

The exposure time required to record the hologram depends on the laser power available, the medium used, and the size and nature of the object(s) to be recorded, as in conventional photography. This determines the stability requirements. Exposure times of several minutes are typical when using powerful gas lasers and silver halide emulsions. All the elements within the optical system must be stable to fractions of a μm over that period. It is possible to make holograms of much less stable objects by using a pulsed laser that produces a large amount of energy in a very short time (μs or less) [1]. These systems have been used to produce holograms of live people. A holographic portrait of Gabor was produced in 1971 using a pulsed ruby laser [2, 3].

Thus, the laser power, recording medium sensitivity, recording time, and mechanical and thermal stability requirements are all interlinked. Generally, the smaller the object is, the more compact the optical layout, so the stability requirements are significantly less than when making holograms of large objects.

Another very important laser parameter is its coherence [4]. This can be envisaged by considering a laser producing a sine wave whose frequency drifts over time; the coherence length can then be the distance over which it maintains a single frequency. This is important because two waves of different frequencies do not produce a stable interference pattern. The coherence length of the laser determines the depth of field that can be recorded in the scene. A good holography laser typically has a coherence length of several meters, which is sufficient for a deep hologram.

The objects that form the scene must, in general, have optically rough surfaces so that they scatter light over a wide range of angles. A specular reflecting (or shiny) surface reflects the light in only one direction at each point on its surface, so in general, most of the light will not be incident on the recording medium. A hologram of a shiny object can be made by locating it very close to the recording plate.

2.4.1 *Experimental Recording of a Hologram*

Consider the experimental arrangement as shown in Fig. 2.1. The laser beam illuminates both objects and behaves as a carrier wave. The carrier wave or reference beam may be a parallel, convergent, or divergent beam. Now, the complex amplitude diffracted from the object over the holographic plane is represented as follows:

$$A_g(x, y) = a_g(x, y) \exp\big[j\phi(x, y)\big]. \tag{2.16}$$

While the complex amplitude of the carrier wave incident directly upon the holographic plate is represented as follows:

$$A_c(x, y) = a_c(x, y) \exp\big[j\psi(x, y)\big]. \tag{2.17}$$

The recorded intensity in the holographic plane (x, y) is:

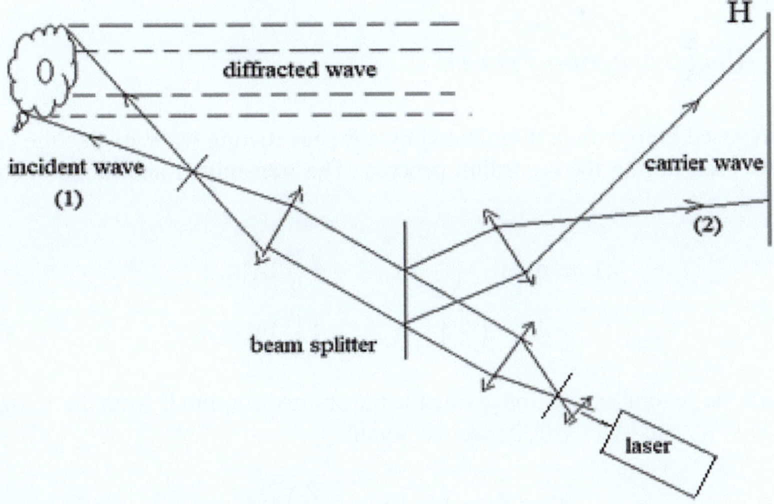

Fig. 2.1 Experimental recording of a hologram

$$I(x, y) = |A_g(x, y) + A_c(x, y)|^2$$
$$= |A_g|^2 + |A_c|^2 + A_g A_c^* + A_g^* A_c. \tag{2.18}$$

Processing the photographic plate under linear conditions, the transmitted amplitude becomes proportional to the recorded intensity, i.e., $t = I^{-\gamma/2}$, γ is the slope of the characteristic curve of the emulsion.

Hence, the transmitted amplitude becomes:

$$t = |A_c|^{-\gamma}\left[1 - \left(\frac{\gamma}{2}\right)\left|\frac{A_g}{A_c}\right|^2 - \left(\frac{\gamma}{2}\right)\frac{A_g A_c^*}{|A_c|^2} - \left(\frac{\gamma}{2}\right)\frac{A_g^* A_c}{|A_c|^2} + \cdots\right]. \tag{2.19}$$

Normalizing the irradiance of the reference beam, i.e., $|A_c|^2 = a_c^2 = 1$, and taking into consideration that $\left(\frac{\gamma}{2}\right)\left|\frac{A_g}{A_c}\right|^2$ is negligible for $A_g < A_c$, then:

$$t = |A_c|^{-\gamma}\left[1 - \left(\frac{\gamma}{2}\right)A_g A_c^* - \left(\frac{\gamma}{2}\right)A_g^* A_c\right]. \tag{2.20}$$

Consequently, we obtain an expression containing both the amplitude and phase of the diffracted wave from the object. Using Eqs. (2.16, 2.17) and substituting in Eq. (2.20), we can rewrite the above expression with the appearance of the interference aspect, as follows:

$$t = \left[1 - \gamma a_g \cos[\phi(x, y) - \psi]\right]. \tag{2.21}$$

2.4.2 Reconstruction Process

The processed hologram is illuminated by a reconstruction wave A_r like the carrier wave A_c used during the recording process. The transmitted amplitude becomes: $A_t = t \cdot A_r$, i.e.,

$$A_t = A_r\left[1 - \left(\frac{\gamma}{2}\right)A_g A_c^* - \left(\frac{\gamma}{2}\right)A_g^* A_c\right]$$
$$= A_r - \left(\frac{\gamma}{2}\right)A_g A_c^* A_r - \left(\frac{\gamma}{2}\right)A_g^* A_c A_r. \tag{2.22}$$

This is the general expression giving the transmitted amplitude from the hologram. Since $A_c = A_r$ and $|A_c|^2 = 1$, hence we obtain:

$$A_t = A_r - \left(\frac{\gamma}{2}\right)A_g - \left(\frac{\gamma}{2}\right)A_r^2 A_g^*. \tag{2.23}$$

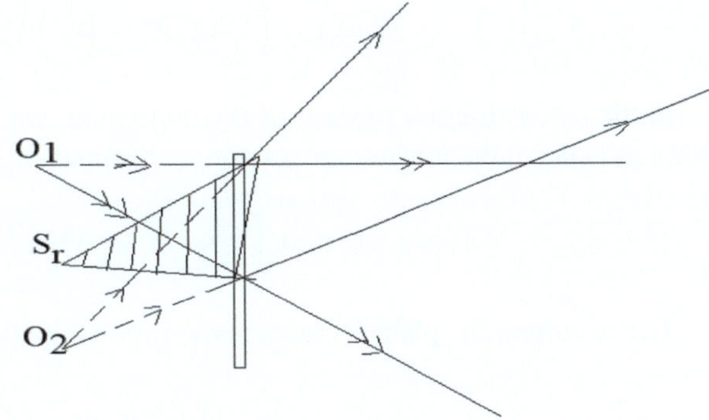

Fig. 2.2 Reconstruction of a hologram. S_r is the reconstruction source

Consequently, the hologram reconstructs the following three waves as in Fig. 2.2:

1. The first term A_r corresponds to the direct transmitted wave. This wave is perturbed by the term $\left(\frac{\gamma}{2}\right)\left|\frac{A_g}{A_c}\right|^2$ which was neglected in Eq. (2.20).
2. Except for the factor $-\frac{\gamma}{2}$, the second term represents the direct image corresponding to the diffracted wave from the object.
3. The third term $-\left(\frac{\gamma}{2}\right)A_r^2 A_g^*$ represents the conjugate image.

2.4.3 Coherent Copy of a Hologram

Since the recording of fringes frequently does not exceed 1000 lines/mm and the fringes constitute a three-dimensional grating, hence it is excluded to utilize optical systems to realize the production of copying. So, the three-dimensional grating structure imposes difficulty in realizing copying by contact. The procedure of coherent copying seems suitable and is summarized as follows: The reconstruction beam that passes through the hologram is considered as a reference beam for the second hologram H' as shown in Fig. 2.3. Hence, the image obtained by H is considered as an object for the hologram H'.

Consider a hologram H, the transmitted complex amplitude is:

Fig. 2.3 Coherent copy of a hologram

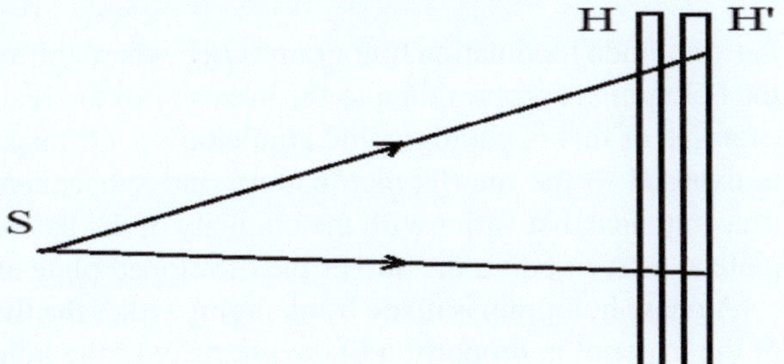

$$t = 1 - \left(\frac{\gamma}{2}\right)A_g A_c^* - \left(\frac{\gamma}{2}\right)A_g^* A_c. \tag{2.24}$$

In the reconstruction process of this hologram, we illuminate it with the carrier wave A_c; hence, the transmitted amplitude becomes:

$$A_t = t \cdot A_c = A_c\left[1 - \left(\frac{\gamma}{2}\right)A_g A_c^* - \left(\frac{\gamma}{2}\right)A_g^* A_c\right]. \tag{2.25}$$

The photographic plate H' records the following intensity:

$$I = |A_t|^2 = A_c^2\left[1 - \left(\frac{\gamma}{2}\right)A_g A_c^* - \left(\frac{\gamma}{2}\right)A_g^* A_c\right]^2. \tag{2.26}$$

The transmitted amplitude from the second hologram H' is:

$$t' = (I)^{-\gamma/2} = \left\{A_c^2\left[1 - \left(\frac{\gamma}{2}\right)A_g A_c^* - \left(\frac{\gamma}{2}\right)A_g^* A_c\right]^2\right\}^{-\gamma/2}$$
$$= A_c^{-\gamma}\left[1 + \gamma\left(\frac{\gamma}{2}\right)A_g A_c^* + \gamma\left(\frac{\gamma}{2}\right)A_g^* A_c\right]. \tag{2.27}$$

Except for the constants $A_c^{-\gamma}$ and γ^2 appeared in Eq. (2.27), the last expression is like Eq. (2.24). Hence, we have obtained an exact copy of the hologram of comparable contrast to the reconstructed image obtained as shown in Fig. 2.3.

2.5 Hologram Classifications

Three important properties of a hologram are defined in this section. A given hologram will have one or more of these three properties, e.g., an amplitude-modulated thin transmission hologram, or a phase-modulated, volume reflection hologram.

2.5.1 Amplitude and Phase Modulation Holograms

An amplitude modulation hologram is one where the amplitude of light diffracted by the hologram is proportional to the intensity of the recorded light. A straightforward example of this is photographic emulsion on a transparent substrate. The emulsion is exposed to the interference pattern and subsequently developed, resulting in a transmittance that varies with the intensity of the pattern—the lighter that fell on the plate at a given point, the darker the developed plate at that point.

A phase hologram is made by changing either the thickness or the refractive index of the material in proportion to the intensity of the holographic interference pattern. This is a phase grating, and it can be shown that when such a plate is illuminated by the original reference beam, it reconstructs the original object wavefront. The efficiency (i.e., the fraction of the illuminated object beam that is converted into the reconstructed object beam) is greater for phase-modulated holograms than for amplitude-modulated holograms.

2.5.2 Thin Holograms and Thick (Volume) Holograms

A thin hologram is one where the thickness of the recording medium is much less than the spacing of the interference fringes that make up the holographic recording. The thickness of a thin hologram can reach 60 nm by using the topological insulator material Sb_2Te_3 thin film [5]. Ultrathin holograms have the potential to be integrated with everyday consumer electronics such as smartphones.

A thick or volume hologram is one where the thickness of the recording medium is greater than the spacing of the interference pattern. The recorded hologram is now a three-dimensional structure, and it can be shown that incident light is diffracted by the grating only at a particular angle, known as the Bragg angle [6]. If the hologram is illuminated with a light source incident at the original reference beam angle but with a broad spectrum of wavelengths, reconstruction occurs only at the wavelength of the original laser used. If the angle of illumination is changed, reconstruction will occur at a different wavelength, and the color of the reconstructed scene will change. A volume hologram effectively acts as a color filter.

2.5.3 Transmission and Reflection Holograms

A transmission hologram is one where the object and reference beams are incident on the recording medium from the same side. In practice, several more mirrors may be used to direct the beams in the required directions.

Normally, transmission holograms can only be reconstructed using a laser or a quasi-monochromatic source, but a particular type of transmission hologram, known as a rainbow hologram, can be viewed with white light.

In a reflection hologram, the object and reference beams are incident on the plate from opposite sides of the plate. The reconstructed object is then viewed from the same side of the plate as that at which the reconstructed beam is incident.

Only volume holograms can be used to make reflection holograms, as only a very low-intensity diffracted beam is reflected by a thin hologram.

2.6 Holographic Recording Media

The recording medium has to convert the original interference pattern into an optical element that modifies either the amplitude or the phase of an incident light beam in proportion to the intensity of the original light field.

The recording medium should be able to fully resolve all the fringes arising from interference between the object and the reference beam. These fringe spacings can range from tens of micrometers to less than one micrometer, i.e., spatial frequencies ranging from a few hundred to several thousand cycles/mm, and ideally, the recording medium should have a response that is flat over this range. Photographic film has a very low or even zero response at the frequencies involved and cannot be used to make a hologram—see, for example, Kodak's professional black and white film whose resolution starts falling off at 20 lines/mm—it is unlikely that any reconstructed beam could be obtained using this film.

If the response is not flat over the range of spatial frequencies in the interference pattern, then the resolution of the reconstructed image may also decrease [7, 8].

The table below shows the principal materials used for holographic recording. Note that these do not include the materials used in the mass replication of an existing hologram, which are discussed in the next section. The resolution limit given in the table indicates the maximal number of interference lines/mm of the gratings. The required exposure, expressed as milli joules (mm J) of photon energy impacting the surface area, is for a long exposure time. Short exposure times (<1/1000 of a second, such as with a pulsed laser) require much higher exposure energies, due to reciprocity failure.

2.7 Image Formation of a Fourier Hologram Using a Carrier Wave [9]

In the following, we present the recording of the Fourier hologram and its reconstruction process.

2.7.1 The Recording of the Fourier Hologram (Coding Process)

A schematic diagram, as shown in Fig. 2.4, is utilized for the realization of the holographic filter. We assume that a plane object represented as $g(x_1, y_1)$ is illuminated with a coherent laser beam of $\lambda = 633$ nm.

In the recording plane H, which is the Fourier transform of the object. Hence, the complex amplitude received in the plane (x_2, y_2) is calculated as follows:

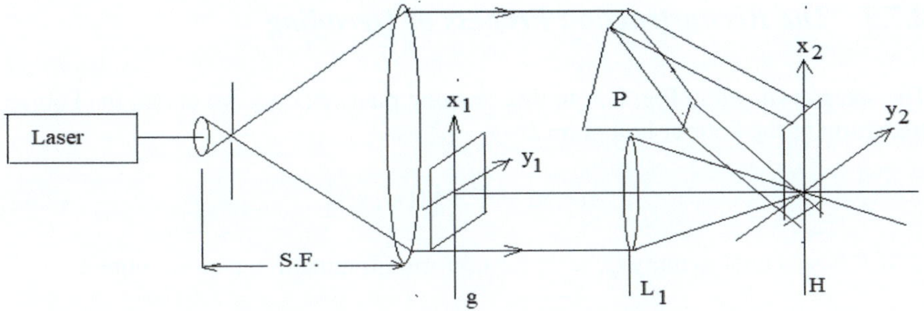

Fig. 2.4 Setup for recording the Fourier hologram of the object $g(x_1, y_1)$

$$H(x_2, y_2) = \text{F.T}\{g(x_1, y_1)\} + \mathcal{R}\exp(-j2\pi \propto x_2) \tag{2.28}$$

where ($\propto$) is the spatial frequency of the reference wave and is given by $\propto = \frac{\sin\theta}{\lambda}$, and θ is the angle between the light propagating along the z-direction and the inclined plane light wave from the prism.

S.F.: spatial filter, P: deflecting prism, L_1: Fourier transform lens, H: holographic plate placed in the Fourier plane (x_2, y_2).

The Fourier transform of the object is calculated as:

$$\tilde{g}\left(\frac{x_2}{\lambda f}, \frac{y_2}{\lambda f}\right) = \int\!\!\!\int_{-\infty}^{+\infty} g(x_1, y_1)\exp\left[\frac{-j2\pi}{\lambda f}(x_1 x_2 + y_1 y_2)\right]dx_1 dy_1. \tag{2.29}$$

It is symbolically written as:

$$\tilde{g}\left(\frac{x_2}{\lambda f}, \frac{y_2}{\lambda f}\right) = \text{F.T}\left[g(x_1, y_1)\right] = \tilde{g}(\zeta, \eta) \tag{2.30}$$

where $\zeta = \frac{x_2}{\lambda f}$, and $\eta = \frac{y_2}{\lambda f}$ are the reduced coordinates in the Fourier plane.

In plane H, we obtain the modulus square of the amplitude $H(x_2, y_2)$ as:

$$I(x_2, y_2) = |H(x_2, y_2)|^2 = \left|\tilde{g}\left(\frac{x_2}{\lambda f}, \frac{y_2}{\lambda f}\right) + \mathcal{R}\exp(-j2\pi \propto x_2)\right|^2$$

$$= \left|\tilde{g}\left(\frac{x_2}{\lambda f}, \frac{y_2}{\lambda f}\right)\right|^2 + |\mathcal{R}|^2$$

$$+ \tilde{g}\left(\frac{x_2}{\lambda f}, \frac{y_2}{\lambda f}\right)\mathcal{R}^*\exp(j2\pi \propto x_2)$$

$$+ \tilde{g}^*\left(\frac{x_2}{\lambda f}, \frac{y_2}{\lambda f}\right)\mathcal{R}\exp(-j2\pi \propto x_2). \tag{2.31}$$

2.7.2 *The Reconstruction Process or Decoding*

The setup is shown in Fig. 2.5. In this imaging plane $P(x, y)$, we obtain the Fourier transform of the Fourier hologram $H(x_2, y_2)$ as:

$$A(x, y) = \text{F.T.}\big[I(x_2, y_2)\big] \tag{2.32}$$

If R is set equal to unity ($R = 1$) for uniform illumination, then we obtain

$$
\begin{aligned}
A(x, y) &= \text{F.T.}\big\{|H(x_2, y_2)|^2\big\} \\
A(x, y) &= \text{F.T.}\{|\tilde{g}|\}^2 + \text{F.T.}\{1\} \\
&\quad + \text{F.T.}\left\{\tilde{g}\left(\frac{x_2}{\lambda f}, \frac{y_2}{\lambda f}\right)\exp(j2\pi \propto x_2)\right\} \\
&\quad + \text{F.T.}\left\{\tilde{g}^*\left(\frac{x_2}{\lambda f}, \frac{y_2}{\lambda f}\right)\exp(-j2\pi \propto x_2)\right\} \\
A(x, y) &= g(x, y) * g^*(-x, -y) + \delta(x, y) \\
&\quad + g(x, y) * \delta(x - f\sin\theta, y) \\
&\quad + g^*(x, y) * \delta(x + f\sin\theta, y) \tag{2.33}
\end{aligned}
$$

Making use of the properties of the convolution with a shifted Dirac-delta function, we finally obtain the following result:

$$
\begin{aligned}
A(x, y) &= g(x, y) * g^*(-x, -y) + \delta(x, y) \\
&\quad + g(x - f\sin\theta, y) + g^*(x + f\sin\theta, y). \tag{2.34}
\end{aligned}
$$

In this text, as shown in Eq. (2.34), the first term represents the autocorrelation of the object $g(x, y) * g^*(-x, -y)$, and the second term represents a point $\delta(x, y)$. These two terms are located at the center, while the third and fourth terms are displaced to

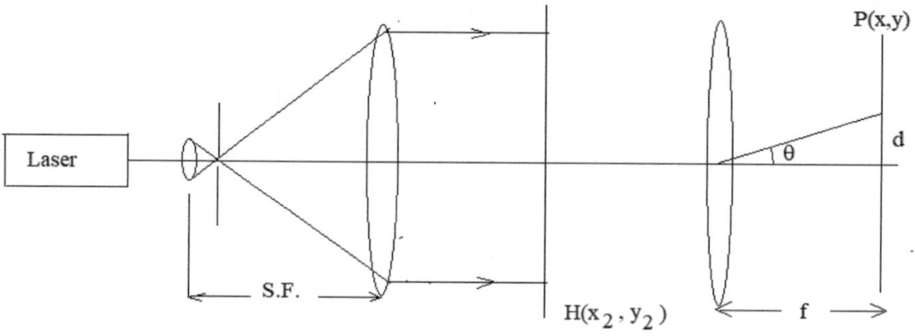

Fig. 2.5 Reconstruction process, where $P(x, y)$ is the imaging plane

the center and are called ± 1 orders located at distances $d = \pm f \sin \theta$. They represent the reconstructed images corresponding to the object.

2.8 Theory of Formation of Image Holograms

2.8.1 Recording of Image Holograms

The holographic recording of a colored plane object is made using a Michelson interferometer. The recording of the colored object as an image hologram is selected for these reasons:

First, the colored components of the object are completely separated while they are not separated in the spectral (focal) plane.

Second, the inclination introduced by the mirror allows the introduction of a separation angle between the object beam and the carrier wave.

Third, the utilization of a small angle of inclination for the reference (carrier) beam doesn't require an emulsion of very high resolution.

Now, assume that a plane polychromatic wave $a(\lambda) \exp\left[j\psi(\lambda)\right]$ created as shown in Fig. 2.6. This polychromatic wave illuminates a simple colored rectangular object composed of only two colors (red and green). The object transmittance is represented as follows:

$$B(x, y, \lambda) = B_1(x, y, \lambda_r) + B_2\left(x, y, \lambda_g\right). \tag{2.35}$$

The object is found along the trajectory $(BS - M_1)$. The carrier wave is created by the reflection from the mirror M_2 with an inclination angle θ. Hence, the complex amplitude recorded in the holographic plane $H(\lambda)$ is:

$$B(x, y, \lambda) = (r_1) \cdot (r) \cdot (t) \cdot a(\lambda) \cdot B(x, y, \lambda)$$
$$+ (r_2) \cdot (r) \cdot (t) \cdot a(\lambda) \exp\left[-j2\pi x \sin(\alpha)/\lambda\right], \tag{2.36}$$

where r, r_1, and r_2 are defined as the amplitude reflection coefficient for the cubic splitter, and the mirrors M_1 and M_2, respectively. t, t_1, and t are defined as the transmission coefficients in amplitude for the cited optical elements. $T = |t|^2$ and $R = |r|^2$ are the intensity transmission and reflection coefficients, and $\alpha = 2\theta$.

The recorded intensity in the imaging plane is obtained by squaring Eq. (2.36):

$$I(x, y, \lambda) = |(r_1) \cdot (r) \cdot (t) \cdot a(\lambda) \cdot B(x, y, \lambda)$$
$$+ (r_2) \cdot (r) \cdot (t) \cdot a(\lambda) \exp\left[-j2\pi x \sin(\alpha)/\lambda\right]|^2.$$

We obtained the intensity as follows:

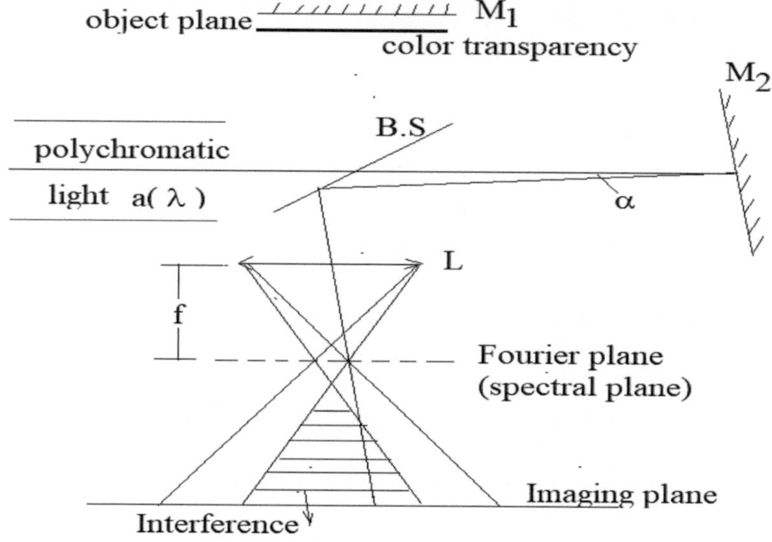

Fig. 2.6 Holographic imaging of a color transparent object

$$I(x, y, \lambda) = a^2(\lambda)RR_1T|B(x, y, \lambda)|^2 + a^2(\lambda)RR_2T$$
$$+ a^2(\lambda)RT\sqrt{R_1R_2}\left\{B^*(x, y, \lambda)\exp\left[-j2\pi x \sin(\alpha)/\lambda\right]\right.$$
$$\left. +B(x, y, \lambda)\exp\left[j2\pi x \sin(\alpha)/\lambda\right]\right\}. \tag{2.37}$$

For a real object let $R_1 = R_2$, we obtained:

$$I(x, y, \lambda) = a^2(\lambda)RR_1T\left[1 + |B(x, y, \lambda)|^2\right]$$
$$+ 2a^2(\lambda)RR_1T\,B(x, y, \lambda)\cos[2\pi x \sin(\alpha)/\lambda]. \tag{2.38}$$

We rewrite Eq. (2.38) as follows:

$$I(x, y, \lambda) = I_0\left\{1 + |B(x, y, \lambda)|^2\right.$$
$$\left. +2B(x, y, \lambda)\cos[2\pi x \sin(\alpha)/\lambda]\right\}$$
$$I_0 = a^2(\lambda)RR_1T. \tag{2.39}$$

It should be noted that the different wavelengths are mutually incoherent, allowing us to write Eq. (2.39) as follows:

$$I(x, y, \lambda) = I_1(x, y, \lambda_r) + I_2(x, y, \lambda_g). \tag{2.40}$$

We assumed two separated color rectangular objects of separation distance $2x_0$ around the center. Hence, we obtained the following:

$$I_1(x, y, \lambda_r) = I_0\big\{1 + |B_1(x - x_0, y, \lambda_r)|^2$$
$$+2B_1(x - x_0, y, \lambda_r)\cos[(2\pi/\lambda_r)x\sin\alpha]\big\}. \qquad (2.41)$$

and

$$I_2(x, y, \lambda_g) = I_0\Big\{1 + |B_2(x + x_0, y, \lambda_g)|^2$$
$$+2B_2(x + x_0, y, \lambda_g)\cos[(2\pi/\lambda_g)x\sin\alpha]\Big\}. \qquad (2.42)$$

Consequently, the holographic recording of a colored object composed of two different colored rectangles gives two holographic gratings in the plane H as in Fig. 2.6.

2.8.2 *Image Reconstruction*

Assume that $P\left(\frac{xf}{d}, \frac{yf}{d}\right)$ to be the pupil that illuminates the hologram H (λ) as shown in Fig. 2.7. Taking into consideration this pupil and considering that the beam is convergent, then the transmitted amplitude is [9]:

$$U(x, y, \lambda) = \sum_i \left(\frac{af}{d}\right)P\left(\frac{x \cdot f}{d}, \frac{y \cdot f}{d}\right)$$
$$\exp\left[-j\left(\frac{k_i}{2d}\right)(x^2 + y^2)\right] \cdot t_i(x, y, \lambda_i) \qquad (2.43)$$

where $t(x, y, \lambda) \sim I(x, y, \lambda)$.

The intensity in the imaging plane is written as:

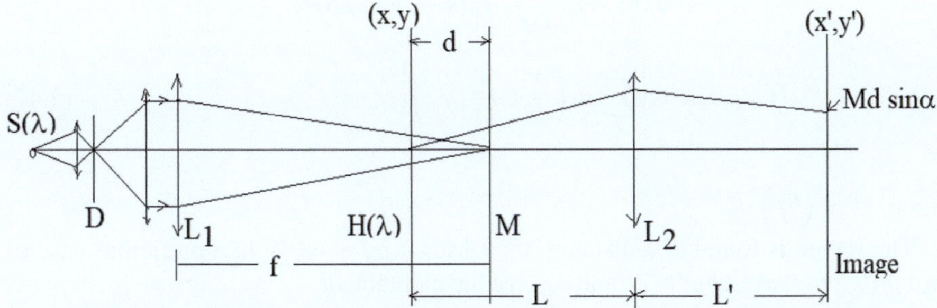

Fig. 2.7 Reconstruction setup using a white light source. *M*: mask placed in the focal plane of the lens *L1*, *D*: diaphragm, *L2*: projecting imaging lens

$$I(x', y', \lambda) = \left| \int\!\!\int_{-\infty}^{\infty} h(x', y'; x, y) U(x, y, \lambda) dx\, dy \right|^2. \qquad (2.44)$$

Substituting Eq. (2.43) in Eq. (2.44), we obtained:

$$I(x', y', \lambda) = \left(\frac{af}{d}\right)^2 \left| \sum_i \int\!\!\int_{-\infty}^{\infty} h(x', y'; x, y) P\left(\frac{x \cdot f}{d}, \frac{y \cdot f}{d}\right) \right.$$

$$\left. \exp\left[-j\left(\frac{k_i}{2d}\right)(x^2 + y^2)\right] \cdot t_i(x, y, \lambda_i) dx\, dy \right|^2 \qquad (2.45)$$

where

$$U = \sum_i U_i(x, y, \lambda_i), \text{ and } t = \sum_i t_i(x, y, \lambda_i) \sim \sum_i I_i(x, y, \lambda_i)$$

$h(x', y'; x, y)$ is the impulse response of the optical system approximately represented by a Dirac-Delta distribution, i.e.,

$$h(x', y'; x, y) = \delta(x' + Mx,\ y' + My) \qquad (2.46)$$

$M = \frac{L'}{L}$ is the magnification of the optical system.

Taking into consideration the incoherence between the different wavelengths, and that the components of the image are spatially separated:

$$I(x', y', \lambda) = C \left| \sum_i B_i\left(\frac{x'}{M}, \frac{y'}{M}, \lambda_i\right) \exp\left[-j\left(\frac{k_i}{2}\right) d \sin^2(\alpha)\right] \right.$$

$$\left. \cdot \exp\left(-j\frac{k_i}{2dM^2}\right)\left[(x' - Md \sin\alpha)^2 + y'^2\right] \right|^2$$

$$= C \left| \sum_i B_i\left(\frac{x'}{M}, \frac{y'}{M}, \lambda_i\right) \right|^2 \qquad (2.47)$$

where $C = \left(\frac{af}{d}\right)^2 I_0$

The image is found at a distance $x' = Md \sin\alpha,\ y' = 0$. In a particular case of two different wavelengths λ_r and λ_g, we have obtained:

$$I(x', y', \lambda) = I_1(x', y', \lambda_r) + I_2(x', y', \lambda_g) \qquad (2.48)$$

where

$$I_1\left(x', y', \lambda_r\right) = C\left|B_1\left(\frac{x'}{M}, \frac{y'}{M}, \lambda_r\right)\right|^2, \text{ and}$$

$$I_2\left(x', y', \lambda_g\right) = C\left|B_2\left(\frac{x'}{M}, \frac{y'}{M}, _g\right)\right|^2.$$

The output efficiency η is:

$$\eta_1 = \frac{I(x', y', \lambda)}{I(x, y, \lambda)} = \frac{C\sum_i\left|B_i\left(\frac{x'}{M}, \frac{y'}{M}, \lambda_i\right)\right|^2}{\sum_i|B_i(x, y, \lambda_i)|^2}. \tag{2.49}$$

For $M = 1$, $\eta_1 = \left(\frac{Af}{d}\right)^2 I_0$, and if $a^2(\lambda) = 1$, $Af = d$, we obtained:

$$\eta_1 = I_0 = RR_1 T.$$

It means that the efficiency of the reconstructed image, neglecting the spectral distribution of the source (uniform distribution), is mainly dependent on the optical properties of the optical components of the interferometer.

The total efficiency is:

$$\eta = \eta_1 \cdot \eta_2$$

where η_2 is the efficiency of the hologram depending upon its type, namely an amplitude hologram or phase hologram.

2.9 Results and Discussion

In this chapter, images with dimensions of 512×512 pixels are used in the formation of the Fourier hologram. The original image of an integrated circuit with dimensions of 256×256 pixels is shown in Fig. 2.8. A diffuser of the same dimensions as the object is shown in Fig. 2.9. The numerical Fourier hologram is computed, as shown in Fig. 2.10, and the reconstructed images are given in Fig. 2.11. Another image of the planet Saturn with dimensions of 320×440 pixels is shown in Fig. 2.12, and the corresponding hologram and its reconstruction are shown in Figs. 2.13 and 2.14.

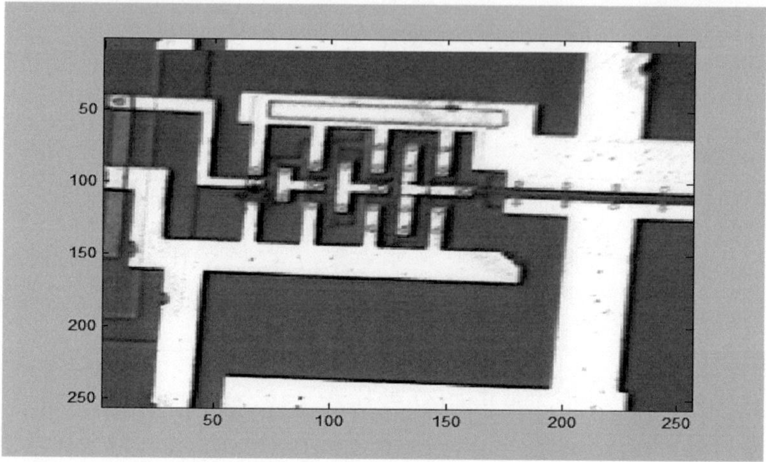

Fig. 2.8 Original image of an integrated circuit with dimensions of 256 × 256 pixels

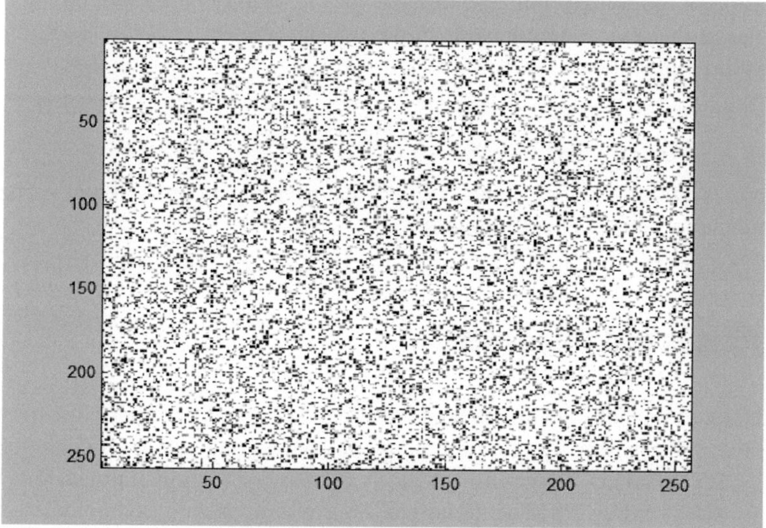

Fig. 2.9 A diffuser with dimensions of 256 × 256 pixels

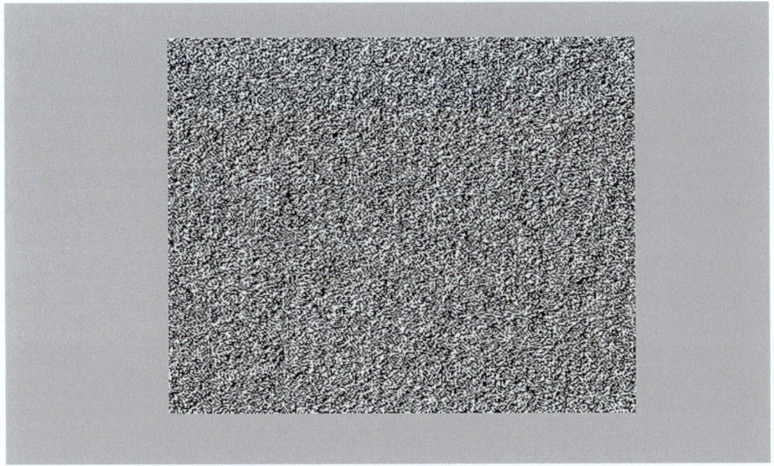

Fig. 2.10   Fourier hologram of the image with dimensions of 256×256 pixels

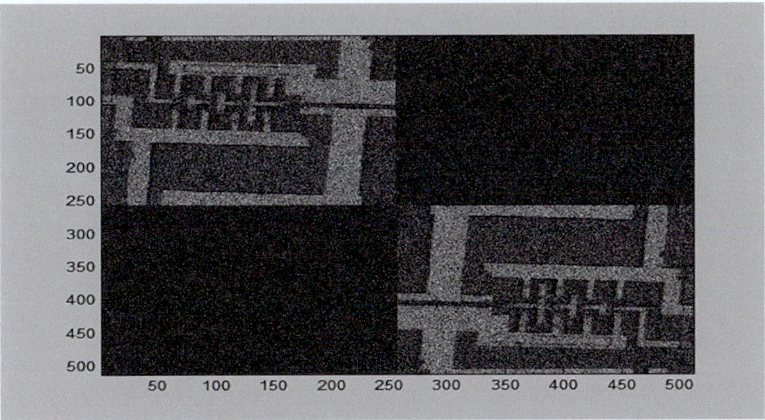

Fig. 2.11   Reconstruction to obtain two conjugate images of dimensions 512×512 pixels

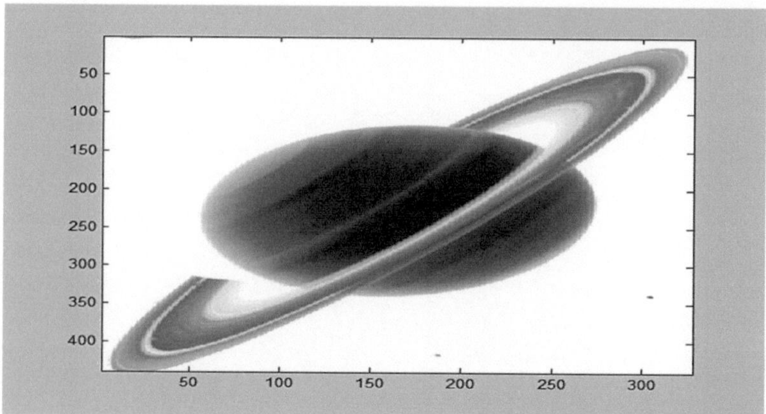

Fig. 2.12 Original image of planet Saturn with dimensions of 320×440 pixels

Fig. 2.13 Fourier hologram of the image with dimensions of 320×440 pixels

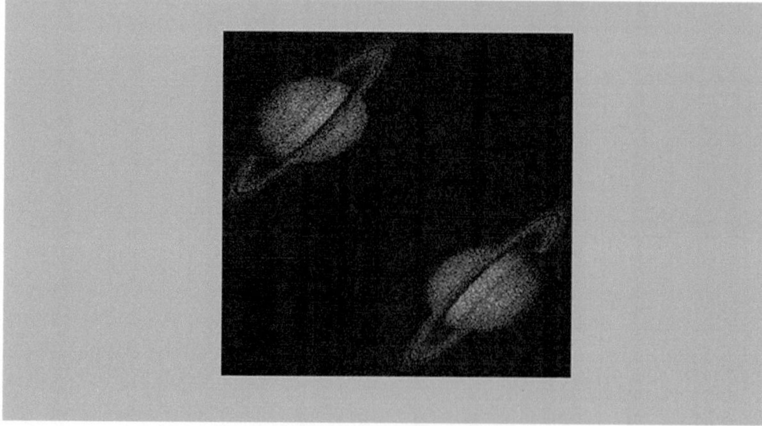

Fig. 2.14 Reconstruction to obtain two conjugated images with dimensions of 640×880 pixels

References

1. D. Gabor, A microscopic new principle. Nature **161**, 777–778 (1948)
2. D. Gabor, Microscopy by reconstructed wavefronts. Proceed. Royal Soc. **197**, 454–487 (1949)
3. M. Francon, *Holography* (Masson, Paris, 1966)
4. W.E. Kock, *Lasers and Holography: An Introduction to Coherent Optics* (Dover Publications, 1981)
5. P. Hariharan, *Optical Holography* (Cambridge University Press, 1996)
6. P. Hariharan, *Basics of Holography* (Cambridge University Press, 2002)
7. A.M. Hamed, Recognition of colored objects using thick holographic multiplexed filter (THMF). Opt. Appl. **13**, 205–213 (1983)
8. H.M. Smith, *Principles of Holography* (Wiley, 1976)
9. J.W. Goodman, *Introduction to Fourier Optics* (McGraw Hill Series, 1996)

Chapter 3
Fourier Holographic Imaging Using Argon Plasma Images

3.1 Introduction

In this chapter, plasma images are processed using Fourier holograms [1–10] for coding and decoding the image. Improved reconstructed images are obtained using Wiener filtering. Profiles of the reconstructed images in both cases before filtering and after filtering are investigated.

3.2 Fourier Holographic Analysis

The plasma image is numerically recorded as a Fourier hologram using MATLAB code as follows: First, a color diffuser of the same dimensions as the rescaled colored image of 512×512 pixels is numerically constructed. Second, the product of both the colored image and the colored diffuser is Fourier transformed numerically to obtain the holographic color interference image. This hologram is the convolution product of the Fourier transformation of the colored image and the colored speckle pattern. The colored speckle distribution is the Fourier transform of the colored diffuser function. Third, the inverse Fourier transform is applied to the described colored hologram to obtain the reconstructed colored images.

The two Fourier transform operations are summarized as follows.

3.2.1 Recording the Amplitude and Phase of the Color Hologram

The complex amplitude of the colored image is analytically represented as:

A. Hamed, *Holographic Imaging Using Aperture Modulation*,
SpringerBriefs in Applied Sciences and Technology,
https://doi.org/10.1007/978-3-031-96989-8_3

$$A(x, y; \lambda) = a(x, y) \exp\left[j\Phi(x, y; \lambda)\right]. \tag{3.1}$$

The colored image is numerically written as follows:

$$A(x, y; \lambda) == \sum_{\lambda_i=1}^{3} \sum_{m=1}^{M} \sum_{n=1}^{N} a(m\Delta x, n\Delta y) \exp\left[j\Phi(m\Delta x, n\Delta y; \lambda_i)\right]. \tag{3.2}$$

where the continuous variables (x, y) are replaced by numerical values as follows:

$$x = m\Delta x \text{ and } y = n\Delta y; \quad j == \sqrt{-1}.$$

The above-colored image is decomposed into three principal components. The monochromatic components are written as follows:

$$A(x, y; \lambda) = A1(x, y; \lambda\text{red}) + A2(x, y; \lambda\text{green})$$
$$+ A3(x, y; \lambda\text{blue}) \tag{3.3}$$

The digital-colored diffuser is written as follows:

$$D(x, y; \lambda) = \sum_{\lambda_i=1}^{3} \sum_{m=1}^{M} \sum_{n=1}^{N} \text{rand}(m\Delta x, n\Delta y) \tag{3.4}$$

where rand is a random value between 0 and 1 and, $\text{rand}(m\Delta x, n\Delta y)$ is the amplitude weighting factor. If the image has square dimensions of 2 cm in height and 2 cm in width and the diffuser has the same dimensions, then $\Delta x = \Delta y = 18$ μm.

The complex amplitude of the Fourier hologram is obtained by operating the FFT upon the multiplication product of the two matrices represented in Eqs. (3.2) and (3.4).

Hence, the complex amplitude of the hologram $B(u, v)$ is represented symbolically as follows:

$$B(u, v; \lambda) = \text{F.T.}\left[A(x, y; \lambda) \cdot D(x, y; \lambda)\right]$$
$$= \text{F.T.}A(x, y; \lambda) \otimes \text{F.T.}D(x, y; \lambda)$$
$$\xrightarrow{\text{yields}} A_{\text{holo}}(u, v; \lambda) \otimes D_{\text{holo}}(u, v; \lambda) \tag{3.5}$$

where (u, v) are the spatial coordinates in the Fourier plane and $\otimes$ is a symbol for convolution.

3.2.2 Reconstruction Process

We apply the inverse Fourier transform upon Eq. (3.5); we obtain:

$$
\begin{aligned}
c(x\prime, y\prime; \lambda) &= \text{F.T}^{-1}[A_{\text{holo}}(u, v; \lambda) \otimes D_{\text{holo}}(u, v; \lambda)] \\
&= A_{\text{reconst}}(x\prime, y\prime; \lambda) \cdot D_{\text{reconst}}(x\prime, y\prime; \lambda)
\end{aligned}
\tag{3.6}
$$

where (x', y') represent the Cartesian coordinates in the imaging plane or the reconstruction plane for the wavelength λ corresponding to the polychromatic illumination. In the case of monochromatic illumination, the colored holographic and reconstructed images are independent of the wavelength, and we obtain Eqs. (3.5) as follows:

$$
B(u, v) = A_{\text{holo}}(u, v) \otimes D_{\text{holo}}(u, v);
\tag{3.7}
$$

$$
c(x\prime, y\prime) = A_{\text{reconst}}(x\prime, y\prime) \cdot D_{\text{reconst}}(x\prime, y\prime).
\tag{3.8}
$$

The holographic image represented by Eq. (3.7), and the reconstructed image represented by Eq. (3.8) are valid for monochromatic illumination. The numerical reconstruction images have maximum dimensions of $x\prime = x_{\max} = M \, \Delta x\prime$ and $y\prime = y_{\max} = N \, \Delta y\prime$ for the matrix of dimensions 1024×1024 pixels.

3.2.3 Computation of the Peak Signal-To-Noise Ratio (PSNR)

We computed the peak signal-to-noise ratio (PSNR) as follows:

We first calculate the mean squared error (MSE) corresponding to the input image $I_1(m, n)$ and the reconstructed image before or after filtration $I_{\text{reconst}}(m, n)$ obtained as follows:

$$
\text{MSE} = \frac{\sum_{m=1}^{M} \sum_{n=1}^{N} [I_1(m, n) - I_{\text{reconst}}(m, n)]^2}{M \times N}
\tag{3.9}
$$

M and N are the number of rows and columns in the input and constructed images. $M = N$ for square matrix. Hence, PSNR is computed from the following formula:

$$
\text{PSNR} = 10 \log_{10}\left(\frac{R^2}{\text{MSE}}\right).
\tag{3.10}
$$

In the previous equation, R is the maximum fluctuation in the input image data type.

3.3   Results and Discussion

The rescaled color plasma image at 512×512 pixels, which is used in polychromatic processing and coding, is shown in Fig. 3.1. A grayscale corresponding image with dimensions of 512×512 pixels is shown on the right side of Fig. 3.1 and is also used for monochromatic processing.

The monochromatic components of the colored image are represented as red in Fig. 3.2a, green in Fig. 3.2b, and blue in Fig. 3.2c. These principal components of colors, namely red, green, and blue, form all colored images with different proportions of colors that are present in the image. The white image is formed from the following combinations as follows:

$$Z = 0.299R + 0.587G + .114B \tag{3.11}$$

where "R" refers to the red spectral component, "G" refers to the green component, and "B" stands for the blue spectral component of the color.

Figure 3.2a–c show that the central region to the right is darker than the other regions, which means that all colors may be present in this region. The colored regions that appeared in all the images showed spectral variation from red through a sharp yellow region to the green band. The band of each colored region is shifted

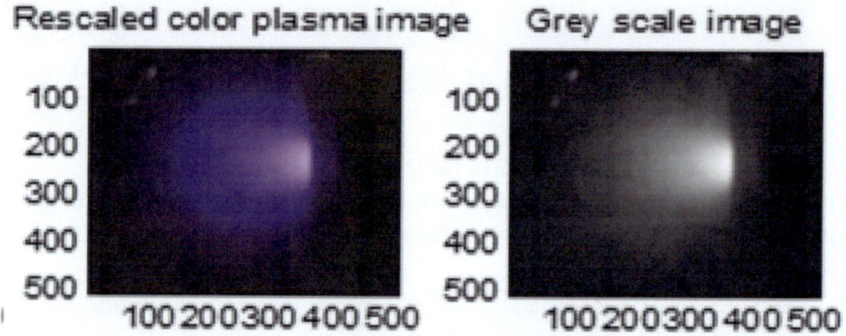

Fig. 3.1   On the left is the rescaled color image with dimensions of 512×512 pixels; and on the right is the grayscale image with dimensions of 512×512 pixels

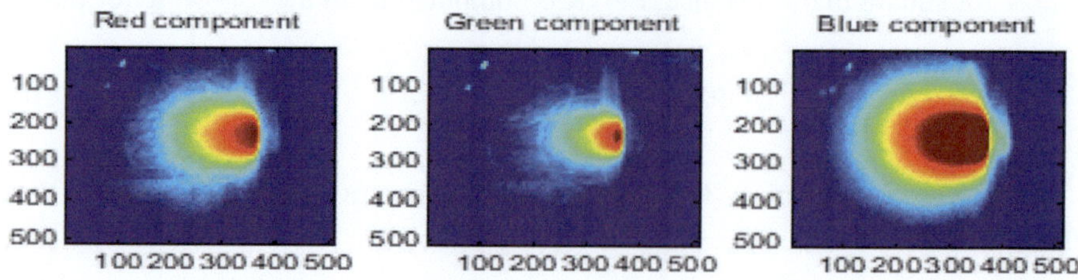

Fig. 3.2   Rescaled plasma images for the red, green, and blue components of the colored image

toward the left for the blue color image, as shown in Fig. 3.2c, which is attributed to the magnification ratio of the wavelengths.

The coding and decoding processes are outlined in Figs. 3.3, 3.4, 3.5, and 3.6. The color diffuser used for the hologram recording is numerically plotted in Fig. 3.3a, and the grayscale diffuser is shown in Fig. 3.3b. Before the Fourier hologram is recorded, the plasma image is coherently multiplied by the complex amplitude of the diffuser, as shown in Fig. 3.4a for the red component of the image and Fig. 3.4b for the grayscale image. Hence, the Fourier hologram is obtained by performing the Fourier transform upon the above multiplication as shown in Fig. 3.5. Figure 2.5a is the Fourier hologram for the red component of the image, while Fig. 3.5b is the hologram for the grayscale image.

The reconstruction process provides two conjugate red images, as shown in Fig. 3.6A, and grayscale images, as shown in Fig. 3.6B. They are obtained by performing the inverse Fourier transform on the complex amplitude of the hologram. It should be noted that the intensity axis is rescaled for all images to 512 ×

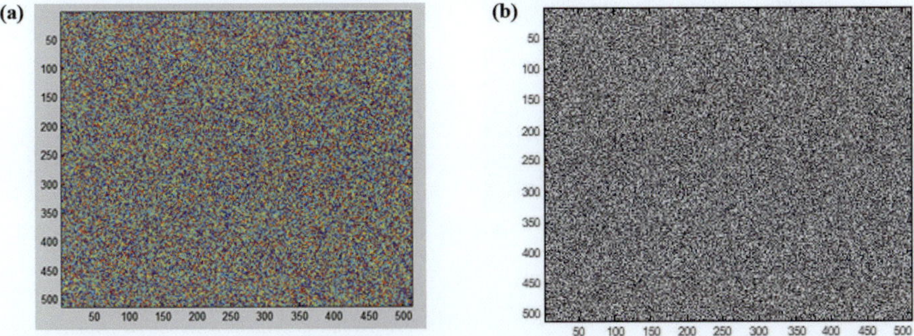

Fig. 3.3 **a** Color diffuser used for hologram recording. **b** Grayscale diffuser used for hologram recording

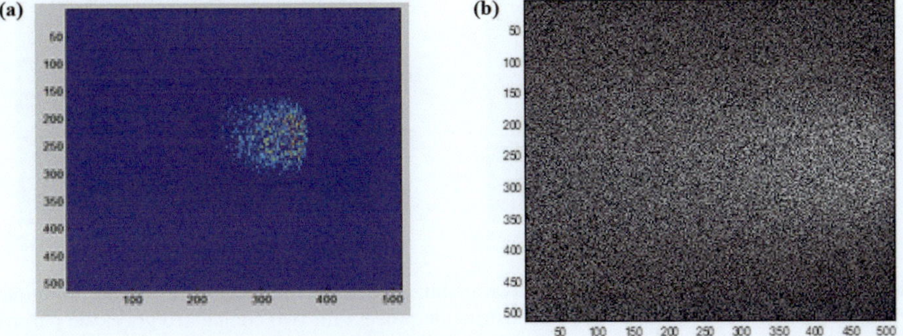

Fig. 3.4 **a** Coherent multiplication of the complex amplitude of the diffuser and the plasma image for the red image shown in Fig. 3.2a. **b** Coherent multiplication of the complex amplitude of the diffuser and the grayscale image shown in Fig. 3.1b

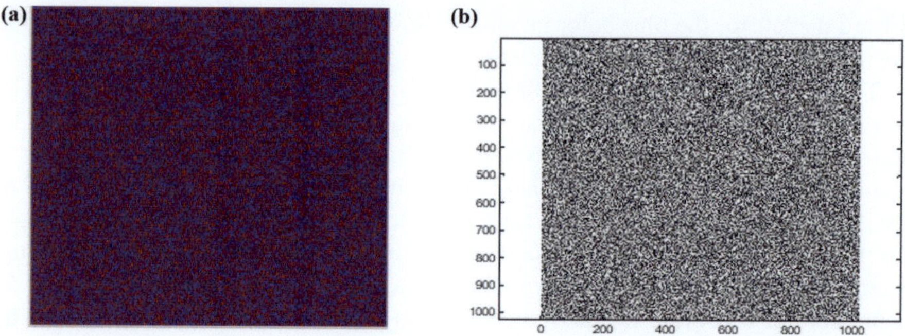

Fig. 3.5 **a** Fourier hologram of the red component of the image with dimensions of 1024×1024 pixels. **b** Fourier hologram of the grayscale image with dimensions of 1024×1024 pixels

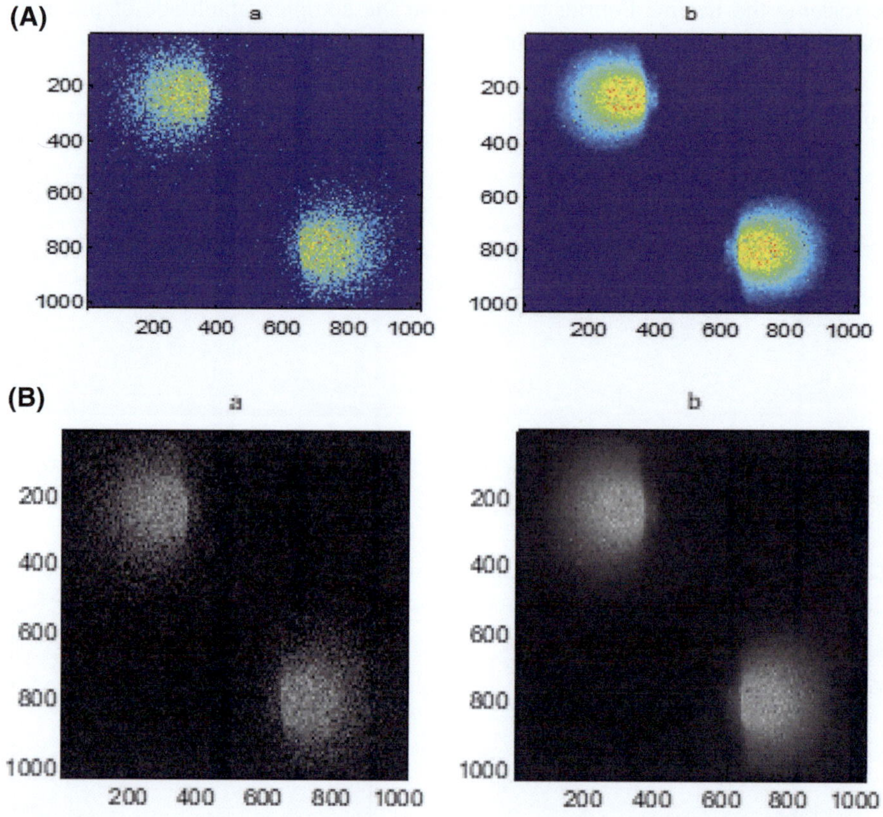

Fig. 3.6 **(A)** Reconstruction process for two conjugate red plasma images. (a) Before filtration, (b) after using Wiener filtration. **(B)** Reconstruction process for two conjugated grayscale plasma images. (c) Before filtration, (d) after filtration

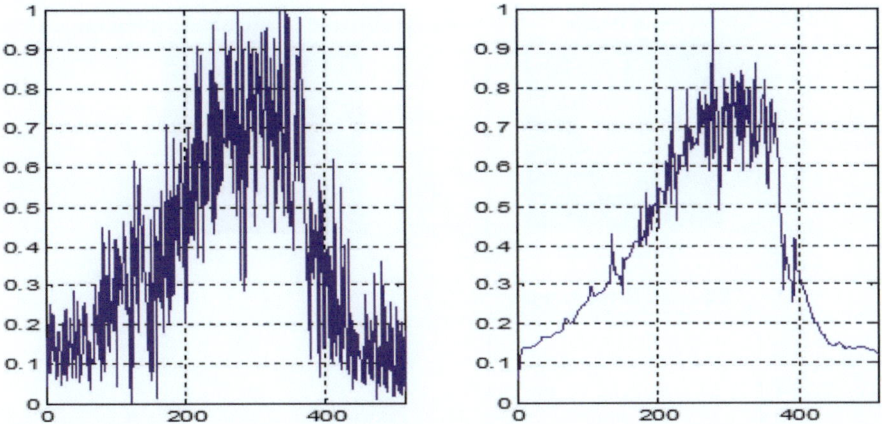

Fig. 3.7 On the left, the profile of the reconstructed image without applying Wiener filters is shown on the left side of Fig. 3.6b. Random noise covers the whole image. On the right, the profile of the reconstructed image after applying Wiener filters is shown on the right side of Fig. 3.6b. The random noise is minimized over the whole image. The two shapes are taken in the same section at 256 pixels in Fig. 3.7

512 pixels, and MATLAB code is written for all processes of plasma image photography in the laboratory by a CCD camera. The conjugate reconstruction images will give a matrix two times the dimension of the hologram matrix, namely 1024 × 1024 pixels. Improved images obtained using a sequence of Wiener filters are shown in Fig. 3.6a and b. The background noise appearing in the reconstructed images is suppressed in the filtered images. Referring to Fig. 3.6b and its corresponding profile shape shown in Fig. 3.7b, the noise is nearly removed from the image in the right image compared with ordinary reconstruction with noise for the image shown on the left.

We obtained the reconstructed image before filtration corresponding to the original plasma image using the Fourier holograms as in Fig. 3.8a. The peak signal-to-noise ratio (PSNR) = 18.04. We obtained reconstructed images after filtration with the Wiener filter shown in Fig. 3.8b. In this case, the peak signal-to-noise ratio (PSNR) = 18.16. It is shown that the PSNR is nearly the same.

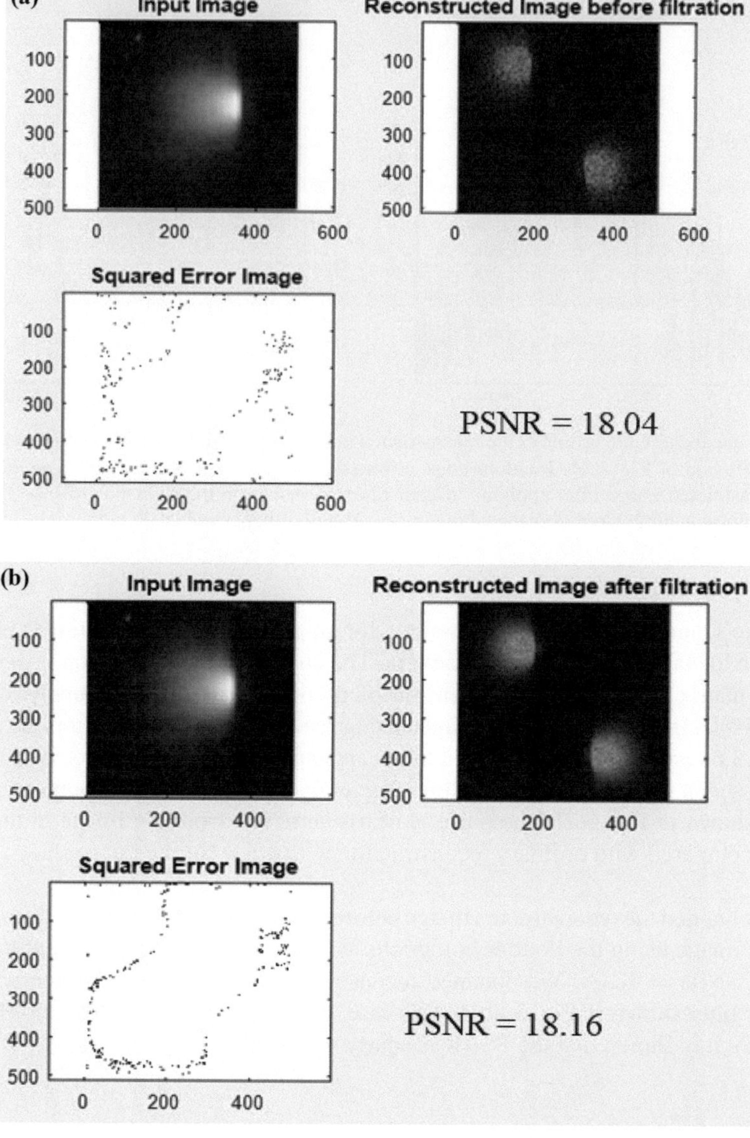

Fig. 3.8 We showed in figure that the PSNR = 18.04 for the reconstructed image before filtration compared to the input grayscale image. After filtration of the reconstructed image, the PSNR = 18.16

3.4 Conclusion

The processing of argon plasma images using Fourier holograms is presented. Wiener filtration is applied to the reconstructed images to improve the images and suppress the background additive noise. The peak signal-to-noise ratio (PSNR) in both cases before and after filtration of the reconstructed images is nearly the same. Additionally, the spectral red, green, and blue basic components of the colored image have distributions in the image proportional to the amount of color present in the image as expected.

References

1. J.W. Goodman, in *Introduction to Fourier Optics and Holography*, 3rd edn. (Roberts & Company Publishers, Greenwood Village, 2005)
2. A.M. Hamed, Polychromatic image processing using thick holographic multiplexed filter. Opt. Appl. **13**, 205–213 (1983)
3. A.M. Hamed, Scanning holography using a modulated linear pupil: simulation. Opt. Photon. J. **1**, 52–58 (2011)
4. A.M. Hamed, M. Saudy, Holographic imaging of argon plasma images. Opt. Photon. J. **4**, 136–142 (2014)
5. A.E. Macgregor, Computer-generated holograms from dot matrix and laser printers. Am. J. Phys. **60**, 839–846 (1992). https://doi.org/10.1119/1.17067
6. S. Trester, Computer simulated holography and computer-generated holograms. Am. J. Phys. **64**, 472–476 (1996). https://doi.org/10.1119/1.18194
7. G. Pedrini, S. Schedin, H.J. Tiziani, Lensless digital holographic interferometry for measurement of large objects. Opt. Commun. **171**, 29–36 (1999). https://doi.org/10.1016/S0030-4018(99)00486-1
8. C. Wagner, S. Seebacher et al., Digital recording and numerical reconstruction of lensless fourier holograms in optical metrology. Appl. Opt. **38**, 4812–4820 (1999). https://doi.org/10.1364/AO.38.004812
9. A.C. Bovik, S.T. Acton, Basic linear filtering with application to image enhancement, in *Handbook of Image and Video Processing*, ed. by A.C. Bovik (Academic Press, San Diego, 2000)
10. M. Agour, E. Kolenovic et al., Suppression of higher diffraction orders and intensity improvement of optically reconstructed holograms from a spatial light modulator. J. Opt. A Pure Appl. Opt. **11**, 105405 (2009). https://doi.org/10.1088/1464-4258/11/10/105405

Chapter 4
Fourier Holographic Images Using Hamming Apertures

4.1 Introduction

All optical microscopes, including conventional widefield, confocal, and two-photon instruments, are limited in the resolution that can be achieved by a series of fundamental physical factors. In a perfect optical system, the resolution is restricted by the numerical aperture of optical components and by the wavelength of light, both incident (excitation) and detected (emission). The concept of resolution is inseparable from contrast, and it is defined as the minimum separation between two points that results in a certain level of contrast between them. In a typical fluorescence microscope, contrast is determined by the number of photons collected from the specimen, the dynamic range of the signal, optical aberrations of the imaging system, and the number of picture elements (pixels) per unit area in the final image [1, 2]. A confocal microscope creates sharp images of a specimen that would otherwise appear blurred when viewed with a conventional microscope. This is achieved by excluding most of the light from the specimen that is not from the microscope's focal plane. The image has less haze and better contrast than that of a conventional microscope and represents a thin cross-section of the specimen [3–6].

The resolution improvement of CLSM over that of conventional optical microscopy is governed by the philosophy of image formation. The resolution of a conventional optical microscope is governed by the resolution limit computed from the ratio between the wavelength of light and the objective numerical aperture. This ratio is computed from the point spread function (PSF). The PSF is obtained from the diffraction pattern of the objective lens located in its focal plane assuming a uniform circular aperture. In contrast, the resolution obtained in the case of CLSM is computed from the RPSF. This RPSF has a narrower pattern than the PSF for conventional optical microscopes. The narrow pattern is due to the multiplication of the two PSFs corresponding to the two conjugate objectives of the confocal arrangement.

© The Author(s), under exclusive license to Springer Nature Switzerland AG 2025

A. Hamed, *Holographic Imaging Using Aperture Modulation*,
SpringerBriefs in Applied Sciences and Technology,
https://doi.org/10.1007/978-3-031-96989-8_4

Many articles on amplitude modulation have improved the resolution of CLSM images early [7–11]. The point spread function (PSF) for an MRC-500 confocal scanning laser microscope using sub-resolution fluorescent beads was measured [12]. PSFs were measured for two lenses with high numerical apertures, which further improved the resolution. The measured PSFs are symmetrical, both radially and axially. Additionally, a theoretical study was presented on a coherent non-scanned microscope (CNSM), which proved its equivalence with a confocal laser scanning microscope (CLSM) [13]. Additionally, confocal laser microscopy has been described as a type of polychromatic image correlation [8].

In medical applications, CLSM can provide real-time, 2D, and 3D images of the cellular morphology and tissue architecture features that pathologists use to detect precancerous lesions without the need for tissue removal, sectioning, or staining [14]. Deconvolution algorithms have proven very effective in conventional (wide-field) fluorescence microscopy. The application of confocal microscopy in biological experiments is hampered by the presence of important levels of noise in the images and by the lack of precise knowledge of the point spread function (PSF) of the system [15].

In this chapter, obstructed and truncated Hamming apertures are applied to CLSM to further improve the resolution compared with uniform circular apertures. Image processing of mammographic photos, using the new truncated Hamming apertures, is applied in CLSM to improve the constructed images. Another application for coding and decoding Hamming apertures using Fourier holograms is presented. Finally, the results and discussion are given, followed by concluding remarks.

4.2   Theoretical Analysis

A confocal scanning laser microscope (CSLM) is used for imaging to improve resolution. This microscope is composed of two symmetric objective lenses where the scanned object is placed in the common short focus of the two objectives, as shown in Fig. 4.1. The constructed image is built in the detector plane point by point during object scanning. Both illumination and detection are coherent in CLSM.

We summarize the known analysis of CLSM in the two Eqs. (4.1 and 4.2). The intensity distribution of the constructed image is computed as the modulus square of the convolution product of the resultant point spread function (RPSF) and the complex amplitude of the object. This intensity is mathematically expressed as follows:

$$I(x, y) = |h_r(x, y) \otimes g(x, y)|^2 \tag{4.1}$$

where $\otimes$ is a symbol for the convolution product.

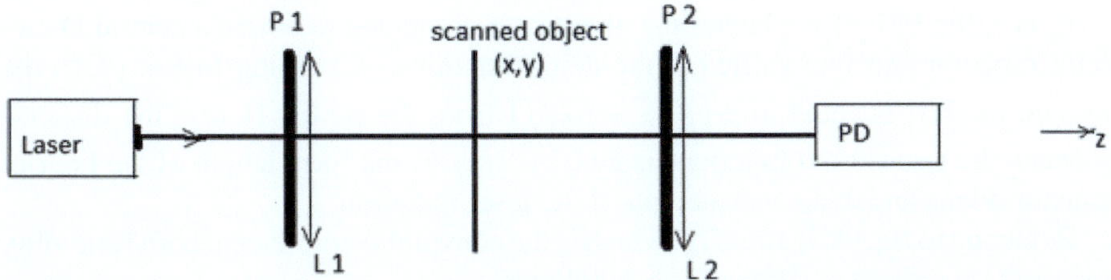

Fig. 4.1 A schematic diagram of the imaging light path in the CLSM. P1 and P2 are the Hamming apertures placed in front of the lenses L1 and L2. A point laser source and a point detector are used in the formation of the image. The image is formed from the convolution product of the point object and the RPSF originated in the common short focus of the lenses L1 and L2. The mechanical scanning of the object synchronized with the electronic scanning is essential for the formation of the image

The RPSF is the multiplication product of the PSF corresponding to each aperture, and the coherent transfer function (CTF) is computed by applying the Fourier transform to the RPSF to obtain the following expression:

$$\mathrm{CTF}(\rho) = \mathrm{F.T}\big[h_r(x, y)\big] = \mathrm{F.T}\big[h_1(x, y) \cdot h_2(x, y)\big]$$
$$= P_1(\rho) \otimes P_2(\rho). \tag{4.2}$$

Now, the obstructed and Hamming Fan apertures proposed by [16] are written as follows:

4.2.1 Central Obstruction of the Hamming Aperture

This obstructed Hamming aperture can be written as follows:

$$P_{\mathrm{ham}}(\rho) = [0.54 + 0.46\cos(\beta\pi(\rho - \rho_2))]; \quad \rho_2 \leq \rho \leq \rho_0$$
$$= 0; 0 < \rho < \rho_2 \tag{4.3}$$

ρ_2: the radius of the central obstructed disk; β is a parameter that has a value between zero and one.

This aperture is Fourier transformed to easily give this expression for the PSF:

$$h_{\mathrm{ham}}(r) = \mathrm{F.T.}\{P_{\mathrm{ham}}(\rho)\}$$
$$= \mathrm{F.T.}[0.54 + 0.46\cos(\beta\pi(\rho - \rho_2))]$$
$$= \left\{\delta(r) + \left(\frac{1}{2}\right)\left[\delta_1\left(r - \beta\lambda\frac{f}{2} - \beta\lambda\rho_2\right)\right.\right.$$
$$\left.\left. + \delta_2\left(r + \beta\lambda\frac{f}{2} + \beta\lambda\rho_2\right)\right]\right\}. \tag{4.4}$$

Hence, the PSF of the Hamming aperture is computed to obtain a central Dirac-delta function and two shifted Dirac-delta functions. According to Eq. (4.4), the amount of shift is equal to $\pm\left(\beta\lambda\frac{f}{2} + \beta\lambda\rho_2\right)$ from the center. Hence, the distance between the two shifted functions is $\beta\lambda(f + 2\rho_2)$. f: the focal length of the Fourier transform lens and λ: the wavelength of the laser radiation.

Referring to Eq. (4.2), the CTF, which is the convolution product of both Hamming apertures, is written in simple form as follows:

$$\begin{aligned}
\text{CTF}_{\text{ham}}(\rho) &= [0.54 + 0.46\cos(\beta\pi(\rho - \rho_2))] \\
&\otimes [0.54 + 0.46\cos(\beta\pi(\rho - \rho_2))]
\end{aligned} \tag{4.5}$$

$\otimes$ is a symbol for convolution.

Equation (4.5) is rewritten in integral form as follows:

$$\begin{aligned}
\text{CTF}_{\text{ham}}(\rho) = \int_{-\infty}^{\infty} &[0.54 + 0.46\cos(\beta\pi(\rho\prime - \rho_2))] \\
&[0.54 + 0.46\cos(\beta\pi(\rho - \rho\prime - \rho_2))]\text{d}\rho\prime.
\end{aligned} \tag{4.6}$$

4.2.2   Truncated Hamming Apertures

The 1st quarter of the Hamming aperture represented by the following equation is written as follows:

$$\begin{aligned}
P_{\text{ham}}(\rho) = [0.54 + &0.46\cos(\beta\pi\rho)]; \\
&0 \leq \rho \leq \rho_0 \text{ and } 0 \leq \theta \leq \pi/2.
\end{aligned} \tag{4.7}$$

The half division of the Hamming aperture is represented by Eq. (4.7), but the azimuthal coordinate is extended to the range $0 \leq \theta \leq \pi$, while three-quarters of the Hamming apertures are in the range $0 \leq \theta \leq 3\pi/2$.

Additionally, the Hamming Fan aperture follows Eq. (4.7), where the azimuthal coordinate θ has the following ranges: in the 1st quarter $0 \leq \theta \leq \pi/2$ and in the 3rd quarter $\pi \leq \theta \leq 3\pi/2$, while in the 2nd and 4th quarters, $\theta = 0$.

4.2.3   Coding and Decoding of the Obstructed and Truncated Hamming Apertures Using Holography

In this section, Fourier holograms of the different obstructed and truncated Hamming apertures are fabricated numerically. The reconstruction process allows the recognition of apertures from the coded holograms.

4.3 Results and Discussion

The mammographic image of the input image as an object is shown in Fig. 4.2. It has dimensions of 1024 × 1024 pixels. In this study, the first model of the CLSM is provided with two identical obstructed Hamming apertures. In the second model, truncated Hamming apertures are used. All aperture photos are in a matrix of dimensions 256 × 256 pixels with a radius of 64 pixels. A comparison with uniform circular apertures is given.

The Hamming aperture in the absence of obstruction ($\rho_2 = 0$), represented by Eq. (4.3), is photographed, and plotted in Fig. 4.3a. The aperture has a total width equal to 128 pixels. This aperture resembles the black-and-white concentric annuli described previously [9]. The difference lies in the presence of the harmonic distribution of illumination found inside the white strip of the Hamming aperture compared with the constant illumination found in the B/W aperture. The obstructed Hamming apertures with two different values of the central black region are plotted in Fig. 4.3b, c. The obstruction's diameters are 32 and 64 pixels, respectively. On the right of the Fig. 4.3a–c, plots of the different obstructions are shown.

The truncated Hamming apertures are used as objects in the formation of Fourier holograms. In Fig. 4.4, in (a), the upper right quarter of the Hamming aperture is shown; in (b), the diffuser used in the recording; in (c), the Fourier hologram of the quarter of the Hamming aperture; and in (d), the reconstructed real and imaginary images are shown. In Fig. 4.5, in (a), the right half of the Hamming aperture is shown; in (b), the diffuser used in the recording; in (c), the Fourier hologram of half of the Hamming aperture; and in (d), the reconstructed real and imaginary images are shown. In Fig. 4.6, in (a), starting from the upper right, three-quarters of the Hamming aperture is shown; in (b), the diffuser used in the recording; in (c), the Fourier hologram of the three-quarters of the Hamming aperture; and in (d), the reconstructed real and imaginary images are shown. In Fig. 4.7, in (a), the Hamming

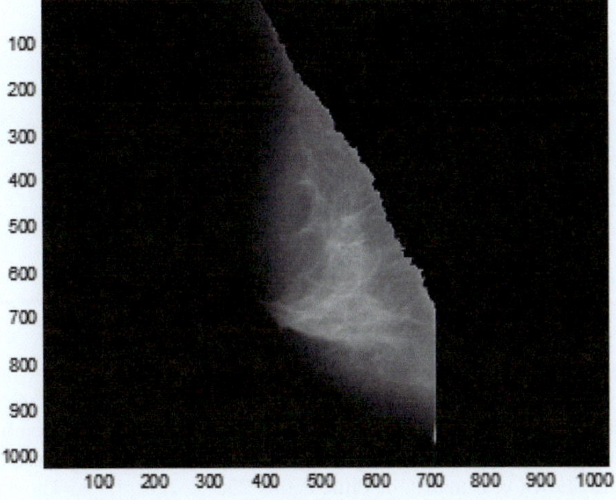

Fig. 4.2 A photo of the mammographic image with dimensions of 1024 × 1024 pixels

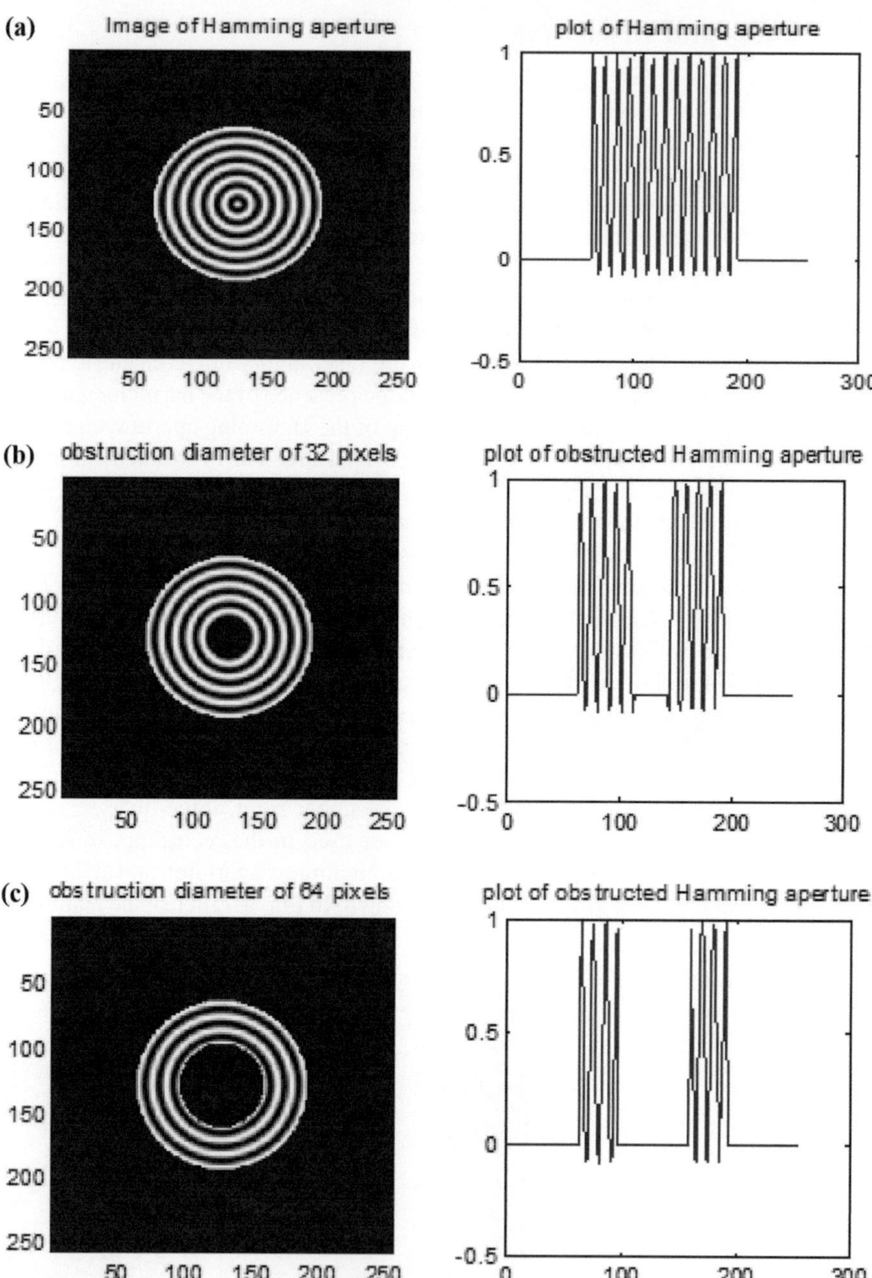

Fig. 4.3 a Image of the Hamming aperture and its plot are shown in a matrix of dimensions 256 × 256 pixels. The total diameter is 128 pixels. **b** Image of the obstructed Hamming aperture and its plot are shown in a matrix of dimensions 256 × 256 pixels. The total diameter is 128 pixels, and the obstruction diameter is 32 pixels. **c** Image of the obstructed Hamming aperture and its plot are shown in a matrix of dimensions 256 × 256 pixels. The total diameter is 128 pixels, and the obstruction diameter is 64 pixels

Fan aperture is shown; in (b), the diffuser used in the recording is shown; in (c), the Fourier hologram of the Hamming Fan aperture is shown; and in (d), the reconstructed real and imaginary images are shown. In all images shown in Figs. 4.4, 4.5, 4.6, and 4.7, the matrix of the object and the diffuser has dimensions of 256×256 pixels, while the matrix of the hologram and the reconstructed images has 512×512 pixels.

The CTF is computed from Eq. (4.6) for the Hamming truncated apertures. Figure 4.8 shows that the CTF has a symmetric ringing shape, which is useful for testing the Hamming aperture. The profiles at different sections at 60, 90, 128, 150, and 180 pixels support this result. Figure 4.9, which shows a segment in the form of a quarter of the Hamming aperture, shows a ringing distribution extending along the quarter and the conjugate. The intensity reaches a maximum at the center at (128, 128) pixels and then decreases along the segment and its conjugate. The CTF shown in Fig. 4.10 for half of the Hamming aperture has a nearly cylindrical ringing shape

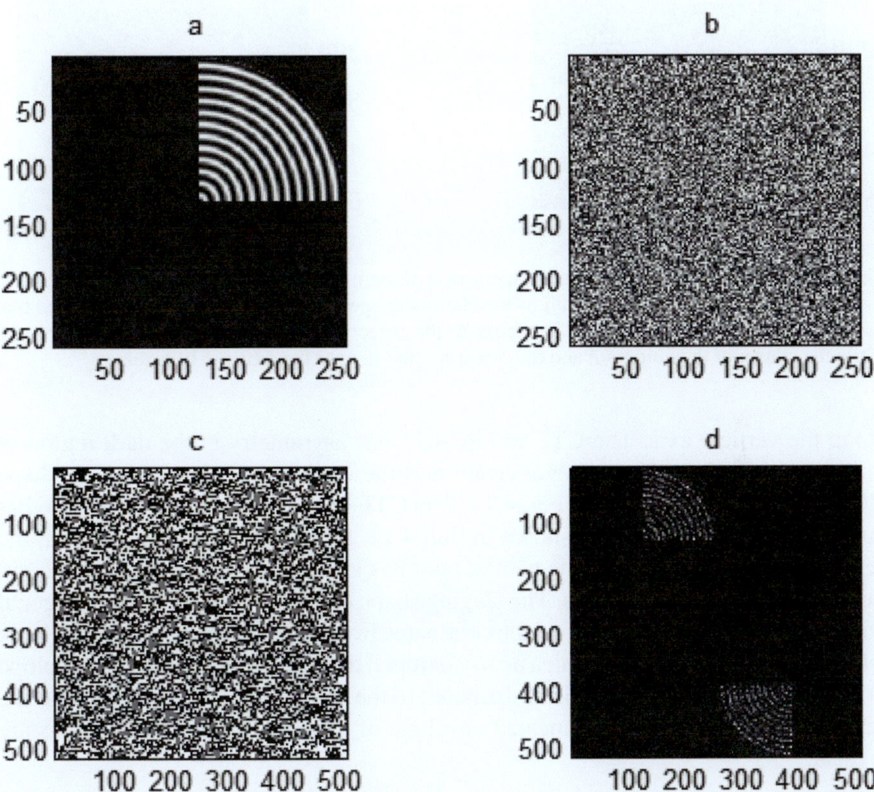

Fig. 4.4 In **a** the upper right quarter of the Hamming aperture is shown; in **b** the diffuser used in the recording; in **c** the Fourier hologram of the quarter of the Hamming aperture; and in **d** the reconstructed real and imaginary images are shown. The matrix of the object and the diffuser has 256×256 pixels, while the matrix of the hologram and the reconstructed images has 512×512 pixels

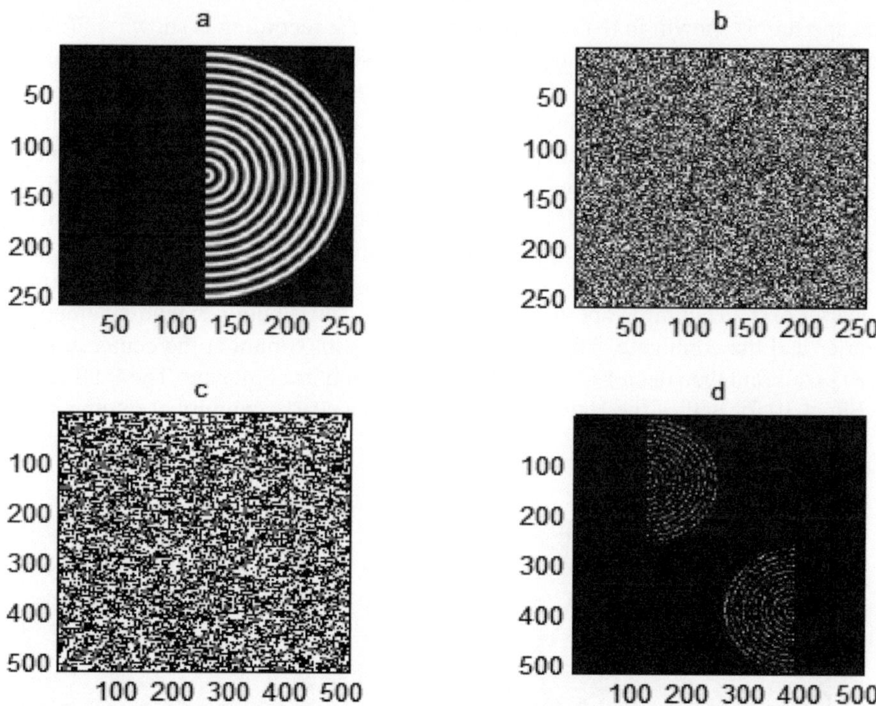

Fig. 4.5 **a** Right half of the Hamming aperture is shown; **b** the diffuser used in the recording is shown; **c** the Fourier hologram of half of the Hamming aperture is shown; **d** the reconstructed real and imaginary images are shown. The matrix of the object and the diffuser has 256×256 pixels, while the matrix of the hologram and the reconstructed images has 512×512 pixels

along the vertical axis. The CTF in Fig. 4.11 has asymmetry in the dark region of three-quarters of the Hamming aperture and the conjugate. It has a ringing shape along the bright part shown in Fig. 4.11. The CTF of the Hamming Fan aperture has quite a new ringing shape, as shown in Fig. 4.12. It has a maximum intensity at the center of (128, 128) pixels. The central spot has an elliptical shape with major and minor axes along the diagonals. The ringing shape oscillates along a certain diagonal with a certain number of rings, which originates from the interference of the aperture wings. Along the other diagonal, a nearly damped intensity with a truncated envelope of wavy shape is localized. This is attributed to the edges of the wings being near the center. Notably, behind the truncated envelope of the wavy shape, dark regions are present.

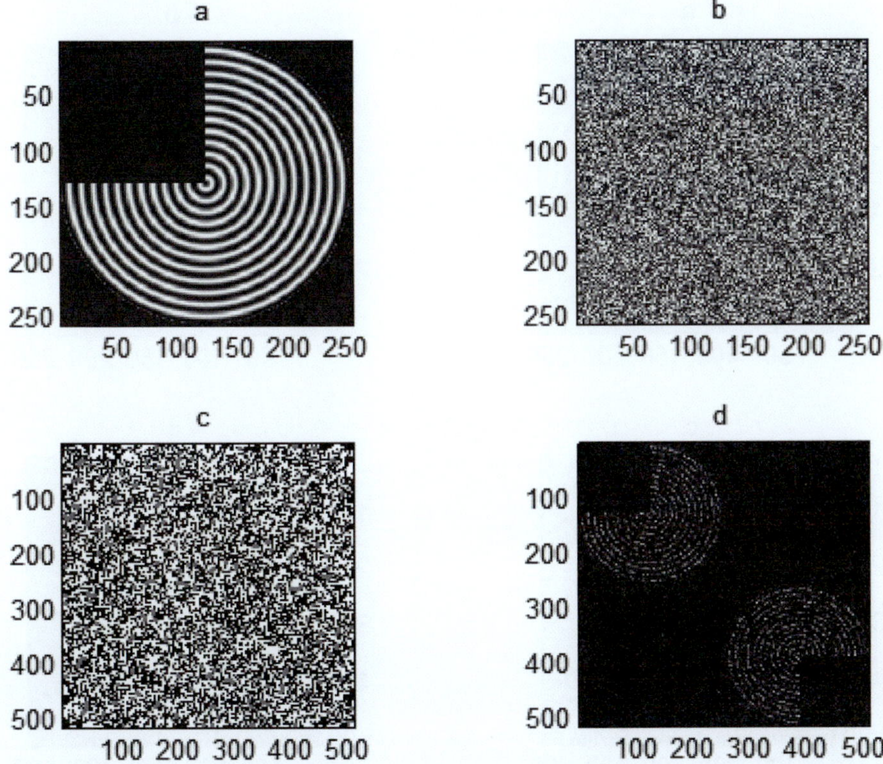

Fig. 4.6 a Starting from the upper right, three-quarters of the Hamming aperture is shown; **b** the diffuser used in the recording; **c** the Fourier hologram of three-quarters of the Hamming aperture; and **d** the reconstructed real and imaginary images. The matrix of the object and the diffuser has 256×256 pixels, while the matrix of the hologram and the reconstructed images has 512×512 pixels

The plot of the PSF represented by Eq. (4.1) using the Hamming aperture is shown in Fig. 4.13. The CTF computed by Eq. (4.2) is shown on the right of Fig. 3.14. The CTF is the convolution product of the symmetric Hamming apertures. According to Figs. 4.13 and 4.14, the PSF has three peaks with one central and two shifted delta functions where the distance of the side peaks from the central peak is set equal to $\pm \beta \lambda \frac{f}{2}$. It is shown from the PSF plotted in Fig. 4.14 that the cutoff spatial frequency is $r_c = 2$ pixels, and the total bandwidth or FWHM is equal to BW $= 4$ pixels. Additionally, the PSF corresponding to the Hamming aperture has an extended distribution, which is useful for imaging extended objects. The CTF has a nearly triangular shape modulated by cosine variations, as shown in Fig. 4.8. It has a total width of 256 pixels compared with the width of the Hamming aperture of 128 pixels.

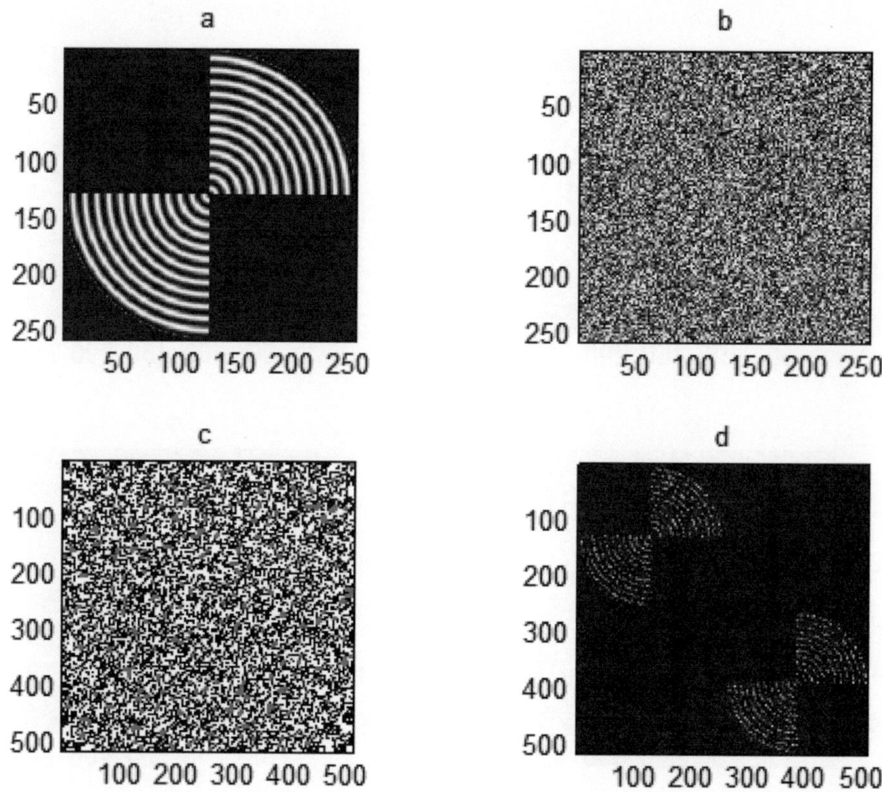

Fig. 4.7 **a** Hamming fan aperture; **b** the diffuser used in the recording; **c** the Fourier hologram of the Hamming fan aperture; **d** the reconstructed real and imaginary images. The matrix of the object and the diffuser has 256×256 pixels, while the matrix of the hologram and the reconstructed images has 512×512 pixels

The PSF for the uniform circular aperture shown in Fig. 4.15 is plotted for comparison with the PSF computed in the case of modulated Hamming apertures. The cutoff values of the PSF spatial frequencies are summarized as follows:

(a)   The cutoff value $r_c = 2$ pixels for the Hamming aperture.
(b)   The cutoff value $r_c = 3.8$ pixels for a uniform circular aperture.

It is shown, from the comparison of the investigated apertures, that the Hamming aperture has a cutoff spatial frequency $r_c = 2$ pixels smaller than the cutoff corresponding to the uniform circular apertures of $r_c = 3.8$ pixels. Hence, the resolution of the Hamming aperture is better than that of the circular aperture.

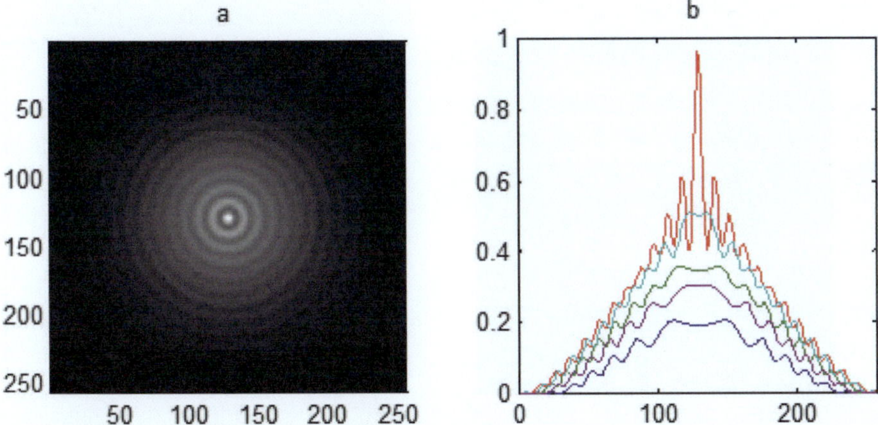

Fig. 4.8 Image of the CTF in a confocal microscope using two symmetric Hamming apertures, each with a diameter $= 128$ pixels, is shown in (**a**). The profile of the CTF at vertical lines at 60, 90, 128, 150, and 180 pixels is shown in (**b**). The total width of the CTF is two times the aperture diameter, which is equal to 256 pixels. The red curve at 128 pixels has the highest intensity, while the blue curve at 60 pixels has the lowest intensity at 0.2

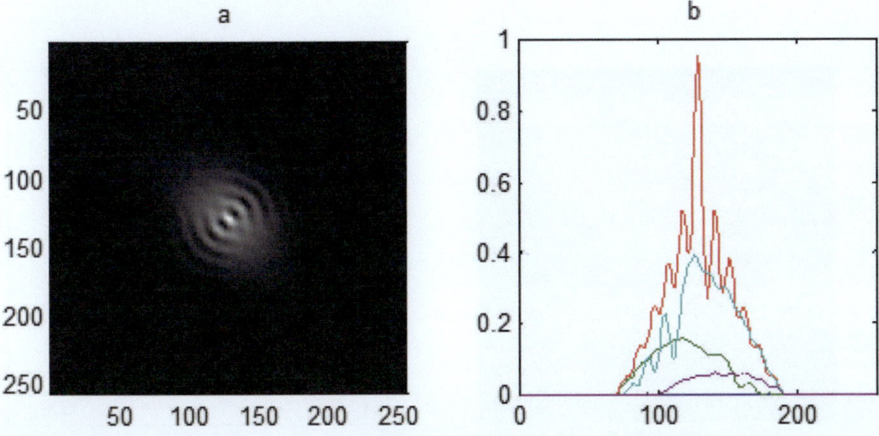

Fig. 4.9 CTF image obtained with a confocal microscope using two-quarters of the Hamming apertures is shown in (**a**). The total width of each aperture $= 64$ pixels. The profile of the CTF at vertical lines at 60, 90, 128, 150, and 180 pixels is shown in (**b**). The blue line at 60 pixels has zero intensity. The total width of the CTF is two times the aperture width, which is equal to 128 pixels. The maximum intensity is located at the center of 128 pixels

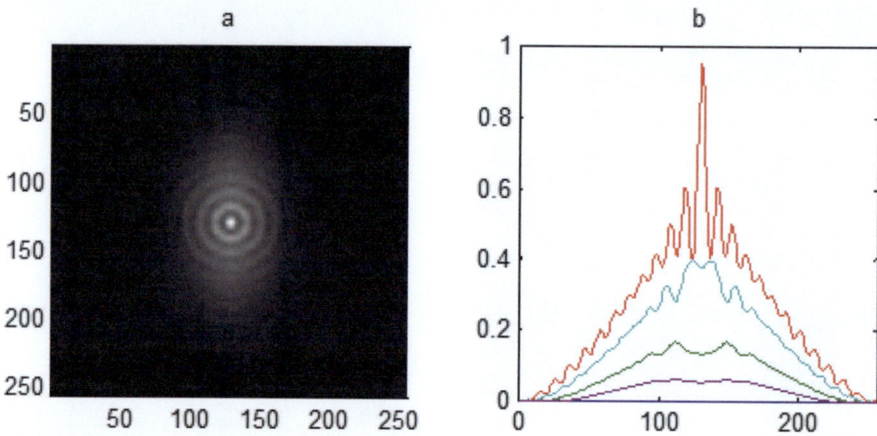

Fig. 4.10 Image of the CTF in a confocal microscope using two symmetric halves of the Hamming apertures is shown in (**a**). The total width of each aperture along the x-axis $= 64$ pixels. The profile of the CTF is shown in (**b**). The total width of the CTF is two times the aperture width, which is equal to 128 pixels. The red curve at 128 pixels has the highest intensity, the green curve at 150 pixels has a lower intensity, the light green curve at 180 pixels, and the violet curve at 90 pixels, while the blue curve has zero intensity at 60 pixels

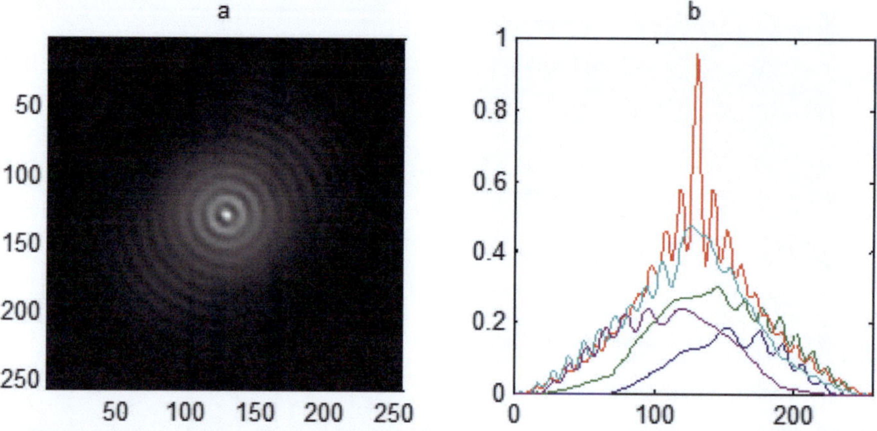

Fig. 4.11 Image of the CTF in a confocal microscope using two symmetric three-quarters of the Hamming apertures is shown in (**a**). The total width of each aperture along the x-axis $= 128$ pixels. The profile of the CTF is shown in (**b**). The total width of the CTF is two times the aperture width, which is equal to 128 pixels. The red curve at 128 pixels has the highest intensity, followed by the green curve at 150 pixels, the light green curve at 180 pixels, and the violet curve at 90 pixels, while the blue curve at 60 pixels has the lowest intensity

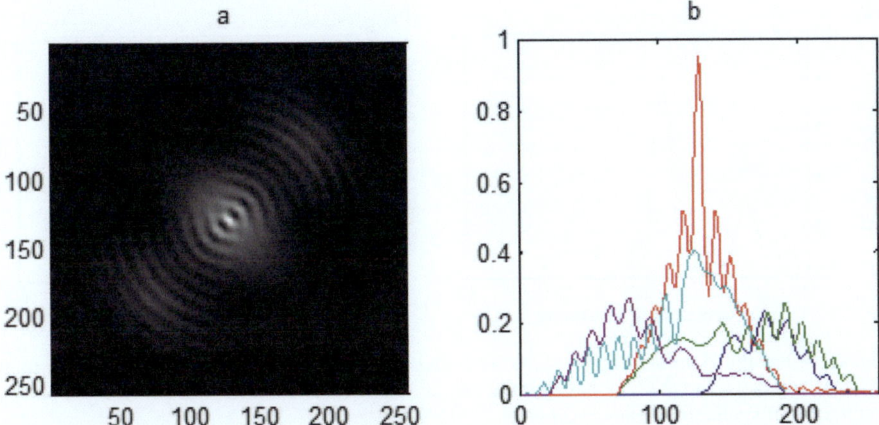

Fig. 4.12 Image of the CTF in a confocal microscope using two symmetric fans of the Hamming apertures is shown in (**a**). The total width of each aperture along the *x*-axis $= 128$ pixels. The profile of the CTF is shown in (**b**). The total width of the CTF is two times the aperture width, which is equal to 128 pixels. The red curve at 128 pixels has the highest intensity at the center, followed by the green curve at 150 pixels, the light green curve at 180 pixels, the violet curve at 90 pixels, and the blue curve at 60 pixels

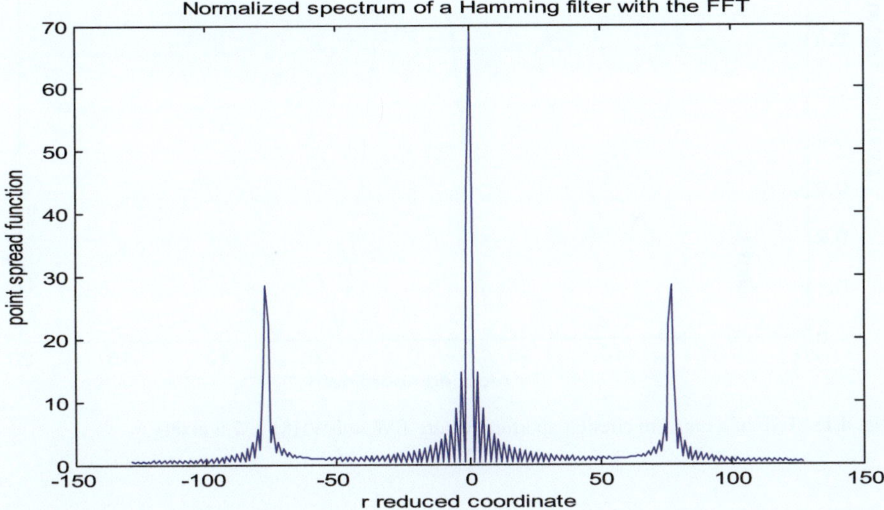

Fig. 4.13 Magnified PSF shape using a matrix of dimensions 256×256 pixels. Two symmetric side peaks are located at 76 pixels around the central peak. The matrix dimensions are 256×256 pixels

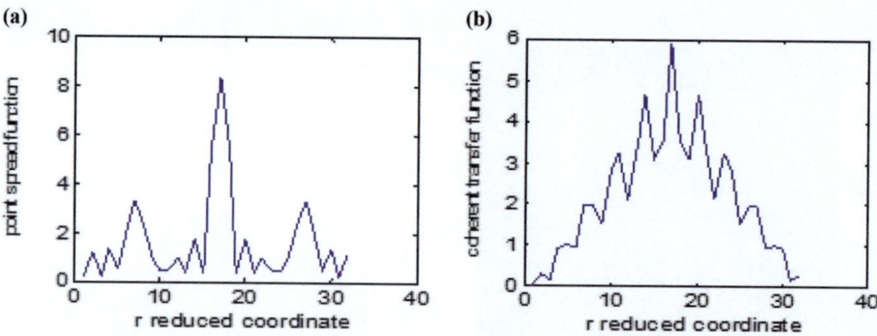

Fig. 4.14 For the Hamming aperture, the PSF and the CTF are shown. The aperture has a total width of 16 pixels, and the CTF has a total width of 32 pixels. The total BW at FWHM = 4 pixels, and the cutoff spatial frequency = 2 pixels

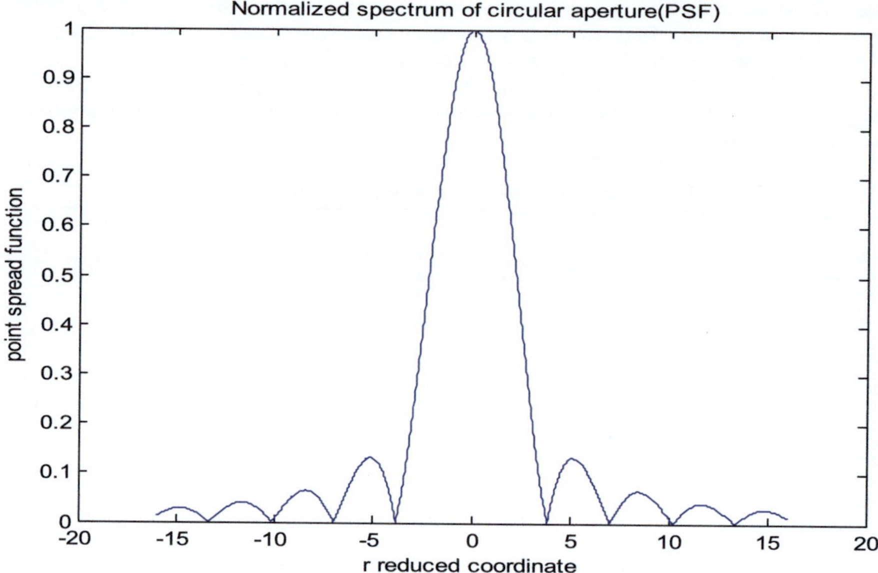

Fig. 4.15 PSF of a uniform circular aperture, where BW = FWHM = 7.6 pixels

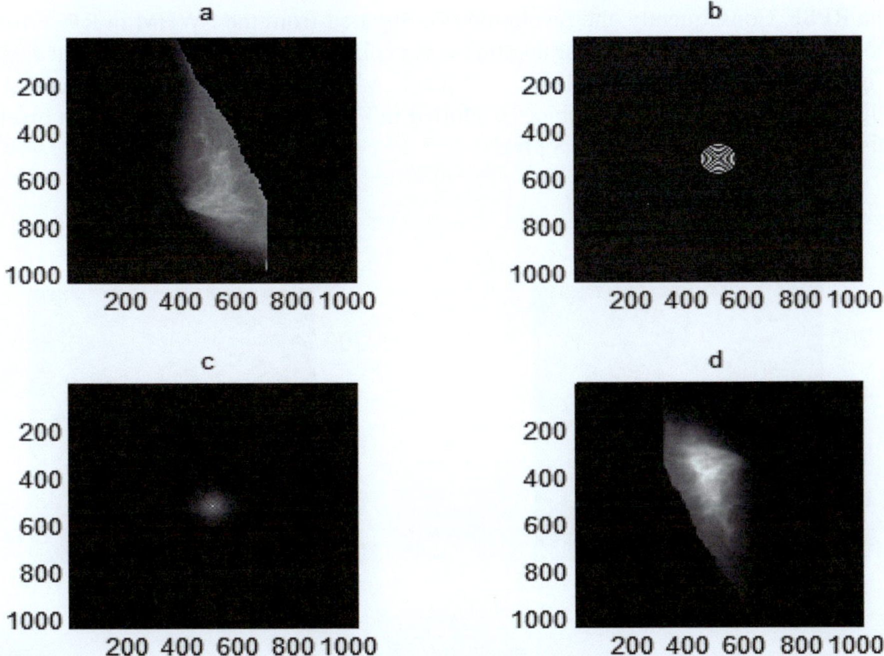

Fig. 4.16   Imaging mammographic object is shown in (**a**), the Hamming aperture of radius 64 pixels is plotted as in (**b**), the coherent transfer function CTF = P1 $\otimes$ P2 is plotted in (**c**), and the image constructed in this confocal coherent laser scanning microscope is computed and plotted as in (**d**). All plots have the same dimensions of 1024 $\times$ 1024 pixels

The image constructed in this coherent laser scanning microscope is computed as follows:

$$I(r) = |h_1 \cdot h_2 \otimes \text{object}(r)|^2$$

where $h_1(r) = \text{F.T.}[P_1(\rho)]$, and $h_2(r) = \text{F.T.}[P_2(\rho)]$. All plots have the same dimensions of 1024 $\times$ 1024 pixels.

The results of the constructed image (d) in CSLM using the Hamming aperture (b) and the corresponding CTF shown in (c) are plotted in Fig. 4.16. The resolution of the constructed image is improved and is dependent upon the RPSF of the imaging system compared with the RPSF obtained in the case of circular apertures shown in Fig. 4.17. This is explained as follows. The image of a point object is equal to the modulus square of the convolution product of the point object represented by the delta function and the RPSF. Hence, the image of a point is the modulus square of the RPSF. Consequently, the resolution is computed from the FWHM in RPSF for both the Hamming and circular apertures, as explained in the results represented by Figs. 4.14 and 4.15. The $FWHM_{ham} = 4$ pixels, while the $FWHM_{circle} = 8$ pixels. A digital image of the original object is plotted in Fig. 4.18. The figure shows a pixel distribution ranging from 1 to 8 pixels.

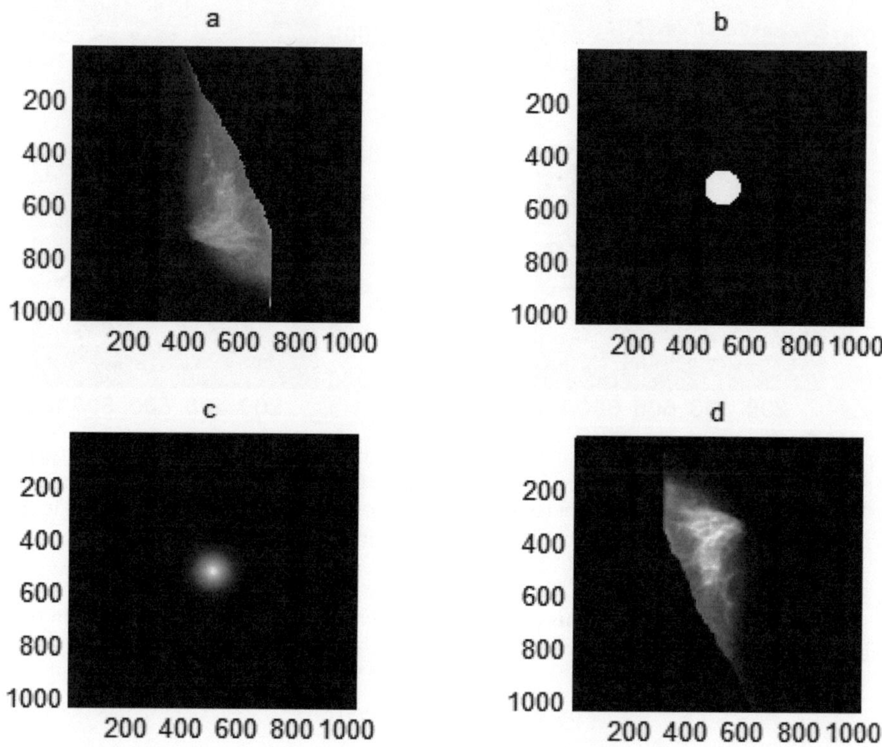

Fig. 4.17 Imaging mammographic object is shown in (**a**), the circular uniform aperture with a radius of 64 pixels is plotted in (**b**), the coherent transfer function CTF $= P1 \otimes P2$ is plotted in (**c**), and the image constructed in this confocal coherent laser scanning microscope is computed and plotted in (**d**). All plots have the same dimensions of 1024 $\times$ 1024 pixels

Fig. 4.18 Digital mammographic image of the malignant image showing that the distribution of pixels varied from 1 to 8 pixels

4.4 Conclusion

First, the comparative results of the different modulated apertures showed that the PSF of the Hamming aperture has three intense peaks. These peaks are considered useful for imaging extended objects. Hence, better resolution is attained in the case of the Hamming aperture compared with the airy disk obtained in the case of the circular aperture.

Second, the CTF of the confocal microscope in the case of truncated Hamming apertures is useful in the recognition of aperture shape modifications. Additionally, the digital construction of the Hamming aperture is considered easier than its fabrication with thin film techniques.

It is shown that the resolution in the case of the Hamming aperture is improved by nearly two times the resolution of the circular aperture. Hence, the image of a point object has a relation to the resolution governed by the measurement of the

FWHM from the PSF. Consequently, the images obtained using a confocal microscope provided with the different Hamming apertures compared with the results obtained in the case of a circular aperture showed better resolution according to the described concept based on the measurement of the FWHM.

The constructed digital mammographic image has a distribution that allows us to predict any abnormalities that appear in the photo. Another application of different Hamming apertures used as objects is the coding and decoding of objects using Fourier holography.

References

1. M. Minsky, Memoir on inventing the confocal microscope. Scanning **10**, 128–138 (1988)
2. D. Semwogerere, R. Eric Weeks, in *Confocal Microscopy* (Emory University, Atlanta, Georgia, 2020)
3. T. Wilson, A.R. Carlini, Three-dimensional imaging in confocal imaging systems with finite-sized detectors. J. Microsc.Microsc. **141**, 51–66 (1988)
4. R.S. Kenneth, T.J. Fellers, W. Michael, *Davidson copy right 2004–2009 OLYMUS Corporation, Florida State University, USA, Resolution and Contrast in Confocal Microscopy* (2024)
5. G. Cox, C.J.R. Sheppard, Practical limits of resolution in confocal and nonlinear microscopy. Microsc. Res. Tech.. Res. Tech. **63**, 18–22 (2004)
6. C.J.R. Sheppard, D.M. Shotton, *Image Formation in the Confocal Laser Scanning Microscope* (Springer, New York, 1997), pp.15–31
7. P. Jansen, R. Hunt, E. Plyler, Resolution enhancement of spectra. J. Opt. Soc. Am. **60**(1970), 596–599 (1970)
8. J.J. Clair, A.M. Hamed, Theoretical studies on optical coherent microscope. Optik **64**, 133–141 (1983)
9. A.M. Hamed, J.J. Clair, Image and super-resolution in optical coherent microscopes. Optik **64**, 277–284 (1983)
10. A.M. Hamed, J.J. Clair, Studies on optical properties of confocal scanning optical microscope using pupils with radially transmission distribution. Optik **65**, 209–218 (1983)
11. A.M. Hamed, Resolution and contrast in confocal optical scanning microscope. Opt. Laser Technol. **16**, 93–96 (1984)
12. P.J. Shaw, D.J. Rawlins, The point-spread function of a confocal microscope: its measurement and use in deconvolution of 3-D data. J. Microsc.Microsc. **163**, 151–165 (1991)
13. A.M. Hamed, Theoretical study on a coherent non-scanned microscope (CNSM). Optik **107**, 89–92 (1998)
14. T. Collier, M. Guillaud, Comparison of two-dimensional and three-dimensional image stacks for detection of cervical precancer. J. Biomed. Opt. **12**, 1154 (2007)
15. J.B. de Monvel, S. Le Calvez, M. Ulfendahl, Image restoration for confocal microscopy: improving the limits of deconvolution, with application to the visualization of the mammalian hearing organ. Biophys. J.. J. **80**, 2455–2470 (2001)
16. A.M. Hamed, T.A. Al-Saeed, Image analysis of modified Hamming aperture: application on confocal microscopy and holography. J. Mod. Opt. **62**, 801–810 (2015). https://doi.org/10.1080/09500340.2015.1007102

Chapter 5
Scanning Holography Using Linear Apertures

5.1 Introduction

Early pioneering work on digital holography or computer-generated holograms (CGHs) was proposed by Goodman and Lawrence [1] and Lohmann and Paris [2], and numerical hologram reconstruction was initiated by Konrad and Yaroslavsky [3] in the early 1970s, followed by many other authors. Improved reconstructed images from CGH are obtained using an iterative operation [4]. Recently, the possibilities of reconstructing the hologram structure and image from a digitally recorded speckle-gram without a reference beam have been considered separately by Hamed [5] and Gorbatenko et al. [6]. Additionally, improved reconstructed images from digital Fourier holograms have been attained using the superposition of reconstructed images obtained at multiple wavelengths [7] and separately using two-step quadratic phase-shifting holography [8], where neither the reference-wave intensity nor an object-wave intensity measurement is needed.

The idea of holographic recording accomplished by heterodyne scanning was originally proposed by Poon [9,10, 11]. Heterodyne scanning was accomplished using a two-pupil optical system by Lohmann and Rhodes [12], who realized Fresnel-zone-plate-type impulse response; i.e., its phase is a quadratic function of x and y, in and out of the focus plane near the focal plane of lenses L_1 and L_2. In the precedent work proposed by Poon, one of the pupils is a delta function, and the other is a constant (uniform circular aperture).

The original idea, which was later analyzed and called scanning holography [13], is to scan a 3D object in a 2D raster with a complex Fresnel-zone plate-type impulse response created by the interference of a point source and a plane wave emerging from each pupil. A temporal frequency offset is introduced between the two pupils, and the desired signal from a spatially integrating detector is obtained via heterodyne detection.

In the present chapter, scanning holographic imaging based on two-pupil heterodyne detection is investigated. In the original standard system proposed by Poon,

© The Author(s), under exclusive license to Springer Nature Switzerland AG 2025

A. Hamed, *Holographic Imaging Using Aperture Modulation*,
SpringerBriefs in Applied Sciences and Technology,
https://doi.org/10.1007/978-3-031-96989-8_5

one of the pupils is a delta function, and the other is a constant. In this chapter, we consider different sets of pupils: one pupil still has a delta function, and the other has a linear distribution. The simulated images reconstructed using the above heterody detection technique are investigated. The proposed linearly modulated aperture [14] was investigated in a recent article on modulated speckle images.

5.2 Theoretical Analysis

The optical scanning hologram is based on two-pupil heterodyne detection, as shown in Fig. 5.1.

5.2.1 Scanning Hologram Using Two Pupils, One Linear and the Other in the Form of the Delta Function

The 1st pupil is chosen to be a linear function distributed within the circular frame of diameter $D = 2\zeta_0$.

$$P_1(x, y) = \zeta; \left|\zeta/\zeta_0\right| \leq 1 \quad \text{for linear aperture.} \tag{5.1}$$

The 2nd pupil remains as before a delta function, which is represented as follows:

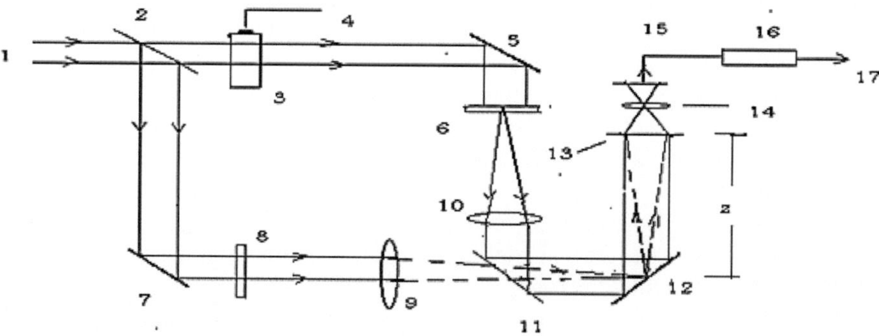

Fig. 5.1 A two-pupil optical heterodyne scanning system. 1—The laser operates at a frequency ω_o. 2, 11—are beam splitters. 3—Acosto-optic frequency shifter. 4—Cos $(\Omega\, t)$ gives a modulated frequency signal at $\omega_o + \Omega$. 5, 7—reflecting mirrors. 2, 5, 7, 11—form the Mach–Zehnder interferometer. 6 and 8—are two pupils, one with a delta function and the other with a linear function. 9, 10—are two converging lenses where the two pupils are located at the front focal planes of lenses L_1 and L_2, both with a focal length of f. 12—A two-dimensional scanning mirror. 13—Object transparency. 14—Collector lens. 15—photodetectors. 16—Electronic bandpass filter tuned at the heterodyne frequency Ω. 17—Output of scanned and processed current $i_\Omega(x, y)$.

$$P_2(x, y) = \delta(x, y). \tag{5.2}$$

The Fourier transform of Eq. (5.1) is computed as follows [15]:

$$P_1(k) = \text{const.}\left[\frac{J_1(k)}{k} + \frac{J_0(k)}{k^2} - \frac{2}{k^2}\sum_i J_i(k)\right]. \tag{5.3}$$

J_o and J_1 are Bessel functions of zero and first orders, respectively. The optical transfer function is obtained as [10]:

$$\text{OTF}_\Omega(k_x, k_y; z) = \exp\left[\frac{jz}{2k_0}\left(k_x^2 + k_y^2\right)\right] \iint P_1^*(x', y')P_2\left(x' + \frac{f}{k_0}k_x, y' + \frac{f}{k_0}k_y\right)$$
$$\exp\left[j\frac{z}{f}(x'k_x + y'k_y)\right]\mathrm{d}x'\mathrm{d}y'. \tag{5.4}$$

In this section, we have assumed a linear function for the 1st pupil and a delta function for the 2nd pupil; hence substituting Eqs. (5.1) and (5.2) into Eq. (5.4), we can write the OTF as follows:

$$\text{OTF}_\Omega(k_x, k_y; z) = \exp\left[\frac{jz}{2k_0}\left(k_x^2 + k_y^2\right)\right] \iint \sqrt{x'^2 + y'^2}$$
$$\delta\left(x' + \frac{f}{k_0}k_x, y' + \frac{f}{k_0}k_y\right)$$
$$\exp\left[j\frac{z}{f}(x'k_x + y'k_y)\right]\mathrm{d}x'\mathrm{d}y'. \tag{5.5}$$

This equation can be rewritten symbolically as follows:

$$\text{OTF}_\Omega(k_x, k_y; z) = \exp\left[\frac{jz}{2k_0}\left(k_x^2 + k_y^2\right)\right]\text{F.T.}\left\{\sqrt{x'^2 + y'^2}\delta\left(x' + \frac{f}{k_0}k_x, y' + \frac{f}{k_0}k_y\right)\right\}. \tag{5.6}$$

Since the Fourier transform of the multiplication product is transformed into a convolution product of the Fourier spectrum of each function [16], Eq. (5.6) becomes:

$$\text{OTF}_\Omega(k_x, k_y; z) = \exp\left[\frac{jz}{2k_0}\left(k_x^2 + k_y^2\right)\right]\text{F.T.}\left\{\sqrt{x'^2 + y'^2}\right\}$$
$$* \text{F.T.}\left\{\delta\left(x' + \frac{f}{k_0}k_x, y' + \frac{f}{k_0}k_y\right)\right\}. \tag{5.7}$$

The Fourier transform of a shifted delta function is calculated to obtain the following result:

$$\text{F.T.}\left\{\delta\left(x' + \frac{f}{k_0}k_x,\, y' + \frac{f}{k_0}k_y\right)\right\} = \exp\left[-\frac{jz}{k_0}\left(k_x^2 + k_y^2\right)\right]. \tag{5.8}$$

By substituting Eq. (5.8) into Eq. (5.7), we obtain:

$$\text{OTF}_\Omega\left(k_x, k_y; z\right) = \exp\left[\frac{jz}{2k_0}\left(k_x^2 + k_y^2\right)\right]\left\{\text{F.T.}\left\{\sqrt{x'^2 + y'^2}\right\}\right.$$
$$\left. * \exp\left[-\frac{jz}{k_0}\left(k_x^2 + k_y^2\right)\right]\right\}. \tag{5.9}$$

The F.T. of the linear function is obtained via Eq. (5.3).

5.2.2 Special Case (Poon Results)

In the case of a uniform circular aperture to represent the 1st pupil instead of the linear aperture, the F.T. becomes:

$$\text{F.T.}\left\{\sqrt{x'^2 + y'^2}\right\} = \text{F.T.}\{1\} = \delta\left(k_x, k_y\right). \tag{5.10}$$

In this case, Eq. (5.9) is reduced to:

$$\text{OTF}_\Omega\left(k_x, k_y; z\right) = \exp\left[\frac{jz}{2k_0}\left(k_x^2 + k_y^2\right)\right]$$
$$\left\{\delta\left(k_x, k_y\right) * \exp\left[-\frac{jz}{k_0}\left(k_x^2 + k_y^2\right)\right]\right\}. \tag{5.11}$$

Since the properties of the convolution product of a function with a delta function remain unchanged, Eq. (5.11) is reduced to the OTF of Poon [13] to yield:

$$\text{OTF}_\Omega\left(k_x, k_y; z\right) = \exp\left[-\frac{jz}{2k_0}\left(k_x^2 + k_y^2\right)\right];\ \text{Poon's result.} \tag{5.12}$$

The intensity distribution of the complex optical scanning hologram, obtained in the case of a uniform circular aperture for the 1st pupil and a delta function for the 2nd pupil, is represented as:

$$H_{c\pm}(x, y) = H_{\cos}(x, y) \pm jH_{\sin}(x, y)$$
$$= \int_D \left\{|\Gamma_0(x, y; z)|^2\right.$$

$$* \frac{jk_0}{2\pi z} \exp\left[\frac{\pm jk_0\left(x^2 + y^2\right)}{2z} \right] \Bigg\} dz. \tag{5.13}$$

where $\Gamma\_0\,(x, y; z)$ represents the complex amplitude corresponding to the object transparency.

$$H_{\cos}(x, y) = \int_D \left\{ |\Gamma_0(x, y; z)|^2 * \frac{k_0}{2\pi z} \cos\left[\frac{k_0\left(x^2 + y^2\right)}{2z} \right] \right\} dz. \tag{5.14}$$

$$H_{\sin}(x, y) = \int_D \left\{ |\Gamma_0(x, y; z)|^2 * \frac{k_0}{2\pi z} \sin\left[\frac{k_0\left(x^2 + y^2\right)}{2z} \right] \right\} dz. \tag{5.15}$$

In the case of the linear pupil combined with the delta function for the 2nd pupil, the intensity distribution of the complex optical scanning hologram is written as follows [17]:

$$H_{c\pm}(x, y) = \int_D \left\{ |\Gamma_0(x, y; z)|^2 * \frac{jk_0}{2\pi z} \exp\left[\frac{\pm jk_0\left(x^2 + y^2\right)}{2z} \right] \right.$$

$$\left. * \left[\sqrt{x^2 + y^2} \cdot \delta\left(x + \frac{f}{k_0}k_x, y + \frac{f}{k_0}k_y\right) \right] \right\} dz \tag{5.16}$$

5.3 Results and Discussion

The original electron microscope image of the reasserted H1N1 influenza virus with dimensions of 180×220 pixels is plotted in Fig. 5.2. The actual dimensions of the image range are 80–120 nm. The autocorrelation intensity of the image is shown in Fig. 5.3.

The reconstructed images are obtained via Fourier transform of the holographic images and are plotted in Fig. 5.4a–f. Reconstruction from the complex holographic FZP images is shown in Fig. 5.4e for constant pupils and in Fig. 4.4f for linear pupils. The reconstructed images from the complex holograms are much better in resolution than the reconstructed images obtained from the cosine and sine holograms. Additionally, the reconstructed FZP image obtained in the case of linear modulation of the 1st pupil, as shown in Fig. 5.4b, is more resolved than the reconstructed image obtained from the sine FZP hologram using a circular uniform pupil Fig. 5.4a. This improvement in image resolution is attributed to the resolution improvement that occurred for the aperture of the linear distribution [14, 15] compared with the constant uniform circular pupils.

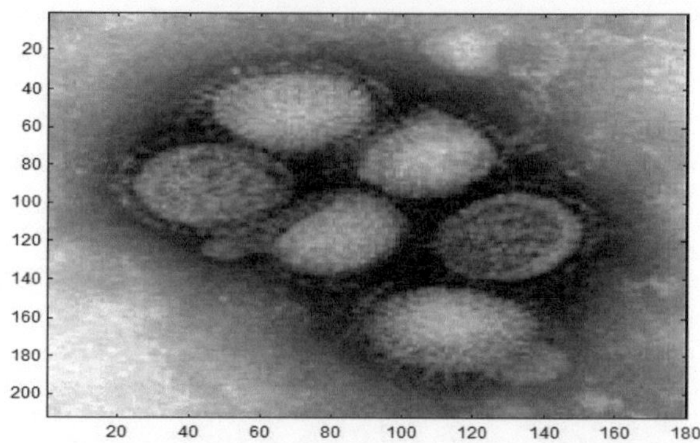

Fig. 5.2 Electron microscopy image of the reasserted H1N1 influenza virus photographed at the CDC influenza laboratory. The viruses are 80–120 nm in diameter. The image has dimensions of 180 × 220 pixels

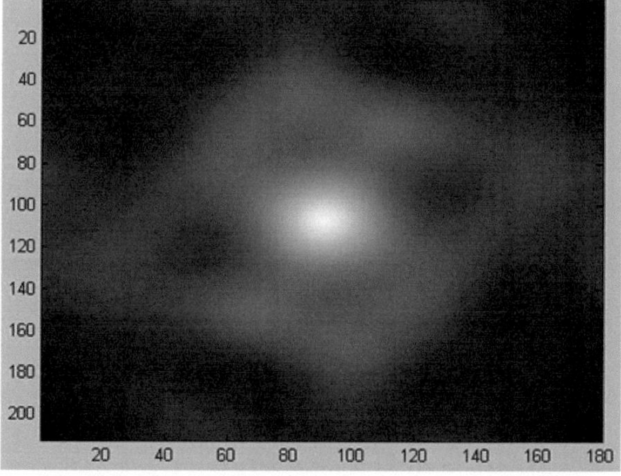

Fig. 5.3 Autocorrelation intensity of the H1N1 image Fig. 4.2. The image has dimensions of 180 × 220 pixels

The image profile of the original image of the H1N1 virus is shown in Fig. 5.5a, the image profile of the sine FZP reconstructed image is shown in Fig. 5.5b, and the image profile of the sine FZP reconstructed image using linear amplitude aperture is shown in Fig. 5.5c. Additionally, the image profile of the reconstructed cosine FZP image is shown in Fig. 5.5d, the image profile of the reconstructed cosine FZP image using linear amplitude modulation is shown in Fig. 5.5e, the image profile of the reconstructed complex FZP image is shown in Fig. 5.5f, and the image profile of the reconstructed complex FZP image using linear amplitude modulation is shown in Fig. 5.5g. All image profiles represented in Fig. 5.5a–g are taken at slice $x = [1\ 212\ 75\ 75]$ and slice $y = [1\ 180\ 100\ 100]$.

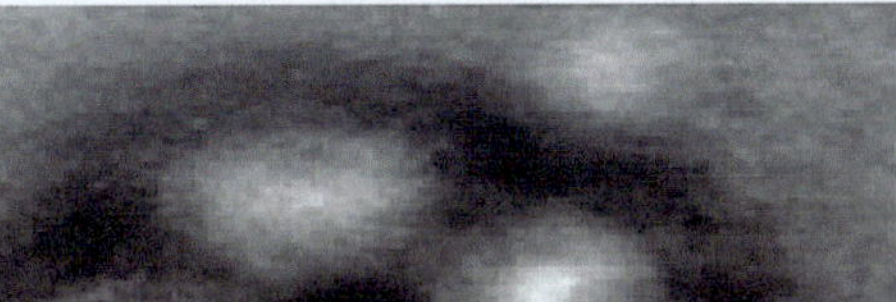

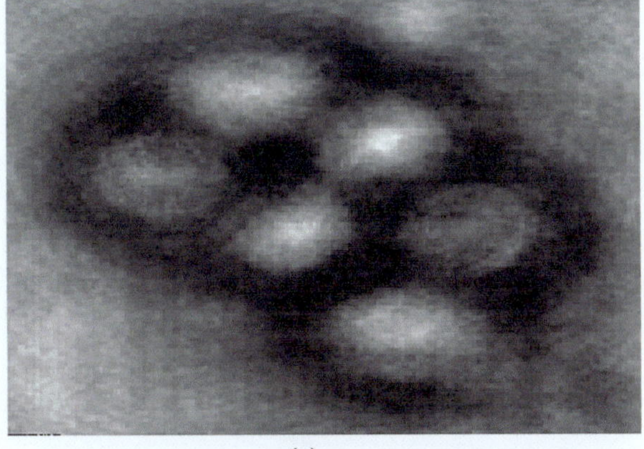

(a)

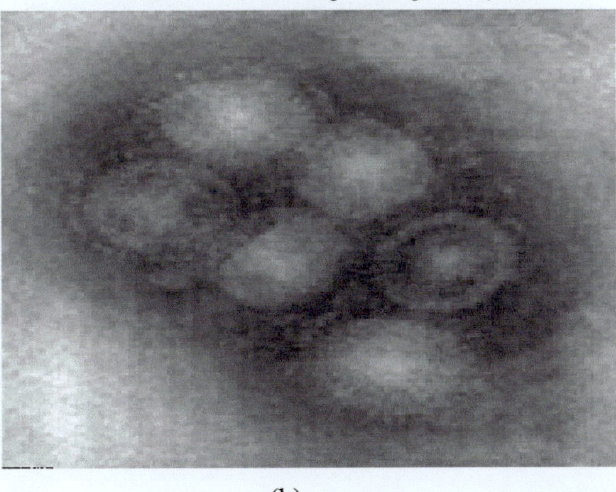

(b)

Fig. 5.4 a Reconstruction of the sine FZP hologram computed using two-pupil heterodyne detection, where the 1st pupil is uniformly circular, and the 2nd pupil is a delta function. **b** Reconstruction of the sine FZP hologram computed using two-pupil heterodyne detection, where the 1st pupil is linearly distributed while the 2nd remains unchanged (delta function). **c** Reconstruction of the cosine FZP hologram computed using two-pupil heterodyne detection, where the 1st pupil is uniformly circular, and the 2nd pupil is a delta function. **d** Reconstruction of the cosine FZP hologram computed using two-pupil heterodyne detection, where the 1st pupil is linearly distributed while the 2nd remains unchanged (delta function). **e** Reconstruction of the complex FZP hologram computed using two-pupil heterodyne detection, where the 1st pupil is uniformly circular, and the 2nd pupil is a delta function. **f** Reconstruction of the complex FZP hologram computed using two-pupil heterodyne detection, where the 1st pupil is linearly distributed while the 2nd remains unchanged (delta function)

Fig. 5.4 (continued)

Reconstruction of cosine-FZP hologram

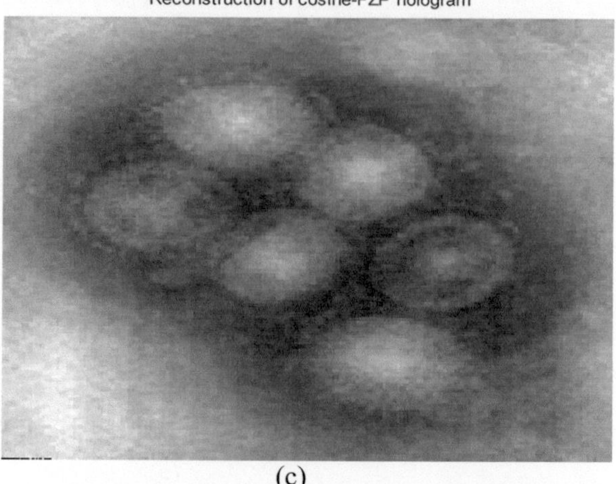

(c)

Reconstruction of cosine-FZP hologram using linear aperture

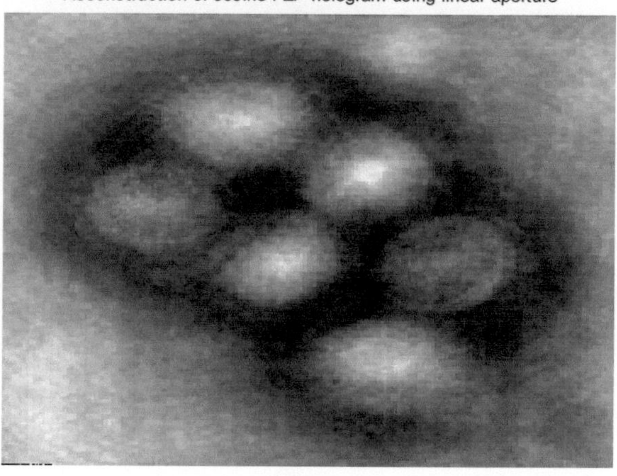

(d)

Fig. 5.4 (continued)

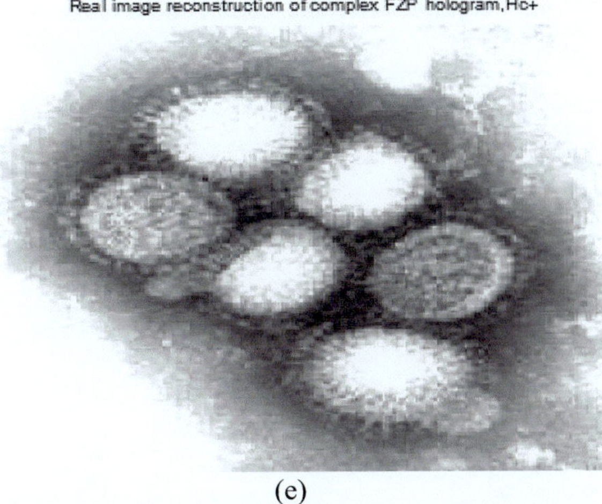

(e)

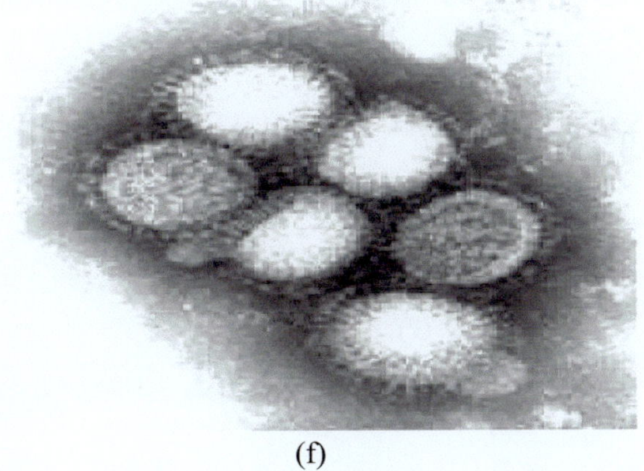

(f)

5.4 Conclusion

First, we conclude that the complex FZP hologram provides better resolution for the reconstructed images than the reconstructed images obtained from the sine and cosine FZP holograms.

Second, the reconstructed images in the case of the sine FZP hologram provided with a linearly modulated aperture are better in resolution than those obtained in the case of the uniform circular pupil.

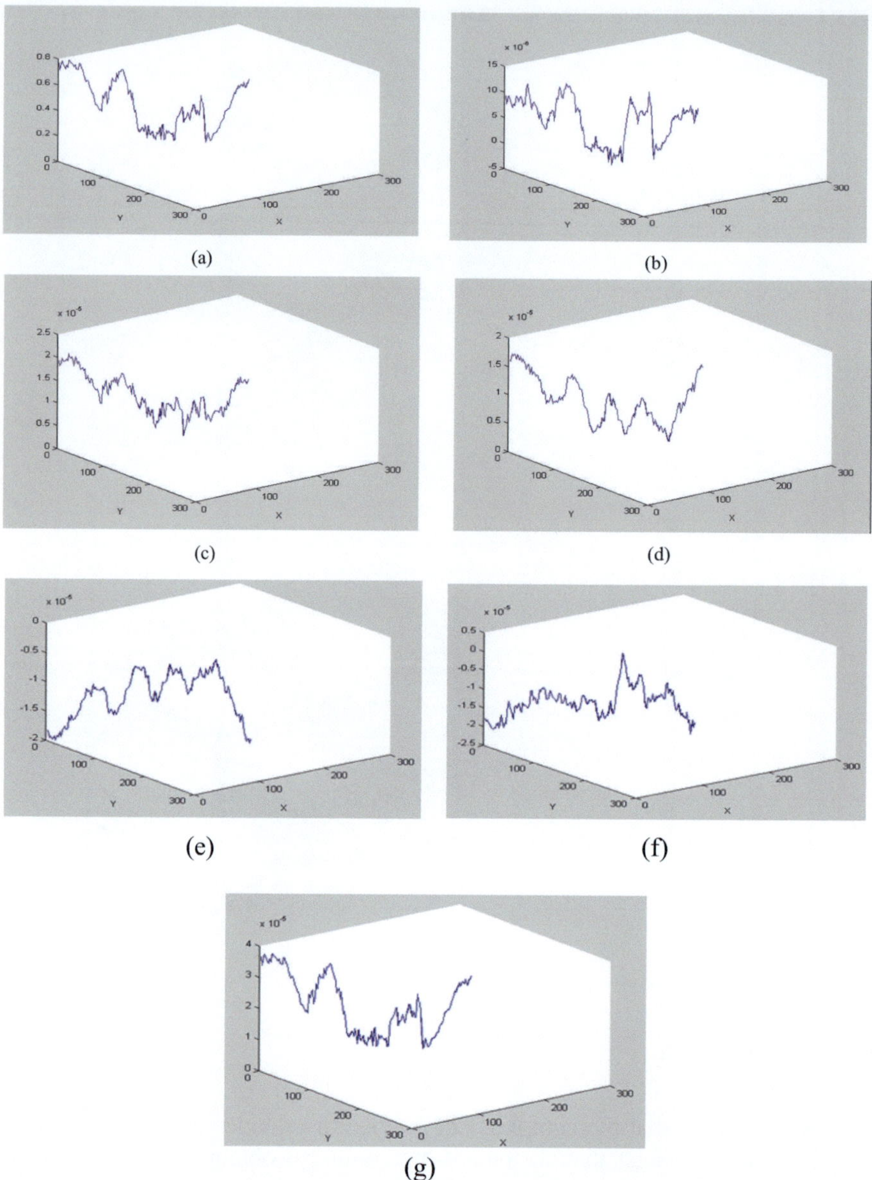

Fig. 5.5 **a** Image profile of the original image of the H1N1 virus. **b** Image profile of the sine FZP reconstructed image. **c** Image profile of the sine FZP reconstructed image using linear amplitude modulation. **d** Image profile of the reconstructed cosine FZP image. **e** Image profile of the reconstructed cosine FZP image using linear amplitude modulation. **f** Image profile of the complex FZP reconstructed image. **g** Image profile of the complex FZP reconstructed image using linear amplitude modulation. All the plots are at slice x [1 212 75 75] and slice y = [1 180 100 100].

References

1. J.W. Goodman, R.W. Lawrence, Digital image formation from electronically detected holograms. Appl. Phys. Lett. **11**, 77–79 (1967)
2. A.W. Lohmann, D.P. Paris, Binary fraunhoffer holograms, generated by computer. Appl. Opt. **6**, 1739–1748 (1967)
3. M.A. Kronrod, L. Yaroslavsky, Computer synthesis of transparency holograms. Sov. Phys. Tech. **17**, 329–332 (1972)
4. K. Nagashima, Improvement of images reconstructed from computer generated holograms using an iterative method. Opt. Laser Tech. **18**, 157–162 (1986)
5. A.M. Hamed, Discrimination between speckle images using diffusers modulated by some deformed apertures: simulations. Opt. Eng. **50**, 1–7 (2011). https://doi.org/10.1117/1.3530085
6. B.B. Gorbatenko, L.A. Maksimova, V.P. Ryabukho, Reconstruction of the hologram structure from a digitally recorded Fourier specklegram. Opt. Spectrosc.Spectrosc. **106**, 281–287 (2009)
7. N. Takanori, O. Mitsukiyo et al., Image quality improvement of digital holography by super-position of reconstructed images obtained by multiple wavelengths. Appl. Opt. **47**, D38–D43 (2008)
8. J.P. Liu, T.C. Poon, Two-step-only quadrature phase-shifting digital holography. Opt. Lett. **34**, 250–252 (2009)
9. T.C. Poon, Scanning holography and two-dimensional image processing by acoustic-optic two-pupil synthesis. J. Opt. Soc. Am. A **4**, 521–527 (1985)
10. T.C. Poon, *Optical Scanning Holography with MATLAB* (2007)
11. T.C. Poon, Recent progress in optical scanning holography. J. Hologr. Speckle **1**, 6–25 (2004)
12. A.W. Lohmann, T. Rhodes, Two-pupil synthesis of optical transfer functions. Appl. Opt. **17**, 1141–1150 (1978)
13. T.C. Poon, T. Kim et al., Twin-image elimination experiments for three-dimensional images in optical scanning holography. Opt. Lett. **25**, 215–217 (2000)
14. A.M. Hamed, Numerical speckle images formed by diffusers using modulated conical and linear apertures. J. Mod. Opt. **56**, 1174–1181 (2009)
15. A.M. Hamed, Image and super-resolution in optical coherent microscopes. Optik **64**, 277–284 (1983)
16. J.W. Goodman, *Introduction to Fourier Optics and Holography* (Roberts and Company Publishers, Englewood, 2005)
17. A.M. Hamed, Scanning holography using a modulated linear pupil: simulations. Opt. Photon. J. **1**, 52–58 (2011)

Chapter 6
Scanning Holography Using Quadratic Apertures

We propose a quadratic aperture for the formation of scanning holographic images to further improve the resolution compared with that of uniform circular and linear apertures. The contrast of the reconstructed image using the quadratic apertures is computed. In this study, the Zika virus image was applied to the scanning hologram. The reconstructed images obtained in the case of scanning holography are investigated and compared with the corresponding images obtained in the uniform circular and linear apertures. The MATLAB code is used in the formation of all the coded and decoded images.

6.1 Introduction

The image of the Zika virus obtained from the transmission electron micrograph shows particles that are 40 nm in size with an outer envelope and an inner dense core. The pioneering work of digital holography or computer-generated holograms (CGHs) was proposed by Goodman et al. [1] and Lohmann et al. [2], and numerical hologram reconstruction was initiated by Kronrod et al. [3] in the early 1970s, followed by many other authors. Improved reconstructed images from CGH are obtained using iterative operations [4]. Recently, the possibilities of reconstructing the hologram structure and image from a digitally recorded speckle-gram without a reference beam have been considered separately by Hamed [5] and Gorbatenko et al. [6]. Additionally, improved reconstructed images from digital Fourier holograms have been attained using superposition of reconstructed images obtained at multiple wavelengths [7] and separately using two-step quadratic phase-shifting holography [8], where neither the reference-wave intensity nor an object-wave intensity measurement is needed in this technique.

© The Author(s), under exclusive license to Springer Nature Switzerland AG 2025

A. Hamed, *Holographic Imaging Using Aperture Modulation*,
SpringerBriefs in Applied Sciences and Technology,
https://doi.org/10.1007/978-3-031-96989-8_6

The idea of holographic recording accomplished by heterodyne scanning was originally proposed by Poon [9–11]. Heterodyne scanning was accomplished using a two-pupil optical system by Lohmann and Rhodes [12]. These authors realized Fresnel-zone-plate-type impulse response, i.e., its phase is a quadratic function of x and y, in and out of the focus plane near the focal plane of lenses L_1 and L_2. In the precedent work proposed by Poon, one of the pupils is a delta function, and the other is a constant (uniform circular aperture).

The original idea was later analyzed and called scanning holography [13]. A temporal frequency offset is introduced between the two pupils, and the desired signal from a spatially integrating detector is obtained via heterodyne detection.

In the present chapter, scanning holographic imaging based on two-pupil heterodyne detection is investigated. In the original standard system proposed by Poon, one of the pupils is a delta function, and the other is a constant, while in this study, the aperture of the quadratic distribution is considered to have a delta function.

A different set of pupils, one with a delta function and the other with a linear function, was proposed for the formation of the scanned holographic image. The proposed linearly modulated aperture [14–16] was investigated in a recent article on modulated speckle images. Fourier transforms and convolution operations are used in the theoretical analysis [17]. Simulated images reconstructed using the above heterodyne detection technique were investigated [18–20].

6.2 Theoretical Analysis

6.2.1 Scanning Hologram Using Two Pupils, One Quadratic, and the Other in the Form of a Delta Function [20]

The optical scanning hologram is based on two-pupil heterodyne detection, as shown in Chap. 5.1, Fig. 5.1. In this chapter, the 1st pupil was chosen to be a circular quadratic distributed function of diameter $D = 2\varrho_0$.

$$P_1(x, y) = \varrho 2; \quad |\varrho/\varrho_0| \leq 1 \quad \text{for quadratic aperture.} \tag{6.1}$$

The 2nd pupil remains as before a delta function, which is represented as follows:

$$P_2(x, y) = \delta(x, y). \tag{6.2}$$

The Fourier transform of Eq. (6.1) is computed by the author in Ref. [16], using integration by parts and recurrence relations. The obtained result is given as follows:

$$h_1(k) = \text{const.} \left[\frac{J_1(k)}{k} - \frac{2J_2(k)}{k^2} \right]. \tag{6.3}$$

J_1 and J_2 are Bessel functions of the first and second orders, respectively. The optical transfer function is obtained as follows [10]:

$$\mathrm{OTF}_\Omega(k_x, k_y; z) = \exp\left[\frac{jz}{2k_0}\left(k_x^2 + k_y^2\right)\right]$$

$$\times \iint P_1^*(x', y')P_2\left(x' + \frac{f}{k_0}k_x, y' + \frac{f}{k_0}k_y\right)\exp\left[j\frac{z}{f}(x'k_x + y'k_y)\right]dx'dy'. \quad (6.4)$$

We assumed a quadratic function for the 1st pupil and a delta function for the 2nd pupil; hence substituting Eqs. (6.1) and (6.2) into Eq. (6.4), we can write the OTF as follows:

$$\mathrm{OTF}_\Omega(k_x, k_y; z) = \exp\left[\frac{jz}{2k_0}\left(k_x^2 + k_y^2\right)\right]$$

$$\times \iint [x'^2 + y'^2]\,\delta\left(x' + \frac{f}{k_0}k_x, y' + \frac{f}{k_0}k_y\right)\exp\left[j\frac{z}{f}(x'k_x + y'k_y)\right]dx'dy'. \quad (6.5)$$

This equation can be rewritten symbolically as follows:

$$\mathrm{OTF}_\Omega(k_x, k_y; z)$$

$$= \exp\left[\frac{jz}{2k_0}\left(k_x^2 + k_y^2\right)\right]\mathrm{F.T.}\left\{[x'^2 + y'^2]\cdot\delta\left(x' + \frac{f}{k_0}k_x, y' + \frac{f}{k_0}k_y\right)\right\}. \quad (6.6)$$

Since the Fourier transform of the multiplication product is transformed into a convolution product of the Fourier spectrum of each function [17], Eq. (6.6) becomes:

$$\mathrm{OTF}_\Omega(k_x, k_y; z) = \exp\left[\frac{jz}{2k_0}\left(k_x^2 + k_y^2\right)\right]\mathrm{F.T.}\{x'^2 + y'^2\}$$

$$\otimes \mathrm{F.T.}\left\{\delta\left(x' + \frac{f}{k_0}k_x, y' + \frac{f}{k_0}k_y\right)\right\}. \quad (6.7)$$

The Fourier transform of a shifted delta function is calculated to obtain the following result:

$$\mathrm{F.T.}\left\{\delta\left(x' + \frac{f}{k_0}k_x, y' + \frac{f}{k_0}k_y\right)\right\} = \exp\left[-\frac{jz}{k_0}\left(k_x^2 + k_y^2\right)\right]. \quad (6.8)$$

By substituting Eq. (6.8) into Eq. (6.7), we obtain:

$$\mathrm{OTF}_\Omega(k_x, k_y; z)$$

$$= \exp\left[\frac{jz}{2k_0}\left(k_x^2 + k_y^2\right)\right]\left\{\mathrm{F.T.}[x'^2 + y'^2]\otimes\exp\left[-\frac{jz}{k_0}\left(k_x^2 + k_y^2\right)\right]\right\}. \quad (6.9)$$

The F.T. of the quadratic function in Eq. (6.9) is obtained in Eq. (6.3), and the OTFΩ becomes:

$$\mathrm{OTF}_\Omega\left(k_x, k_y; z\right) = \exp\left[\frac{jz}{2k_0}\left(k_x^2 + k_y^2\right)\right]$$
$$\times\left\{\left[\frac{J_1(k)}{k} - \frac{2J_2(k)}{k^2}\right] \otimes \exp\left[-\frac{jz}{k_0}\left(k_x^2 + k_y^2\right)\right]\right\}. \tag{6.10}$$

6.2.2   Special Case (Poon Results)

The intensity distribution of the complex optical scanning hologram, obtained in the case of a uniform circular aperture for the 1st pupil and a delta function for the 2nd pupil, is summarized as follows:

$$Hc_\pm(x, y) = H_{\cos}(x, y) \pm jH_{\sin}(x, y)$$
$$= \int_D \left\{|\Gamma_0(x, y; z)|^2 \otimes \frac{jk_0}{2\pi z}\exp\left[\frac{\pm jk_0(x^2 + y^2)}{2z}\right]\right\}dz \tag{6.11}$$

$$H_{\cos}(x, y) = \int_D \left\{|\Gamma_0(x, y; z)|^2 \otimes \frac{k_0}{2\pi z}\cos\left[\frac{k_0(x^2 + y^2)}{2z}\right]\right\}dz \tag{6.12}$$

$$H_{\sin}(x, y) = \int_D \left\{|\Gamma_0(x, y; z)|^2 \otimes \frac{k_0}{2\pi z}\sin\left[\frac{k_0(x^2 + y^2)}{2z}\right]\right\}dz. \tag{6.13}$$

In the case of the quadratic pupil combined with the delta function for the 2nd pupil, the intensity distribution of the complex optical scanning hologram is written as follows:

$$Hc_\pm(x, y) = \int_D \left\{|\Gamma_0(x, y; z)|^2 \otimes \frac{jk_0}{2\pi z}\exp\left[\frac{\pm jk_0(x^2 + y^2)}{2z}\right] \otimes \left[(x^2 + y^2)\right.\right.$$
$$\left.\left.\delta\left(x + \frac{f}{k_0}k_x, y + \frac{f}{k_0}k_y\right)\right]\right\}dz. \tag{6.14}$$

6.3 Results and Discussion

The original image with dimensions of 652 × 434 pixels resized to dimensions of 256 × 256 pixels is plotted in Fig. 6.1. The actual diameter is nearly 40 nm.

Images of the apertures used in the formation of the scanning hologram from the left are circular, linear, and quadratic apertures in Fig. 6.2. The images have dimensions of 256 × 256 pixels and an aperture radius = 64 pixels.

The PSF corresponding to the circular, linear, and quadratic apertures is computed from the FFT algorithm and plotted as shown in Fig. 6.3a. In addition, it is computed from Eq. (6.3) for the quadratic aperture and compared with the PSF for the circular aperture, as shown in Fig. 6.3b. From the curves in Fig. 6.3b, it is shown that the quadratic aperture has a cutoff value less than that corresponding to the circular aperture. Consequently, the quadratic aperture resolution is better than the resolution of the circular aperture.

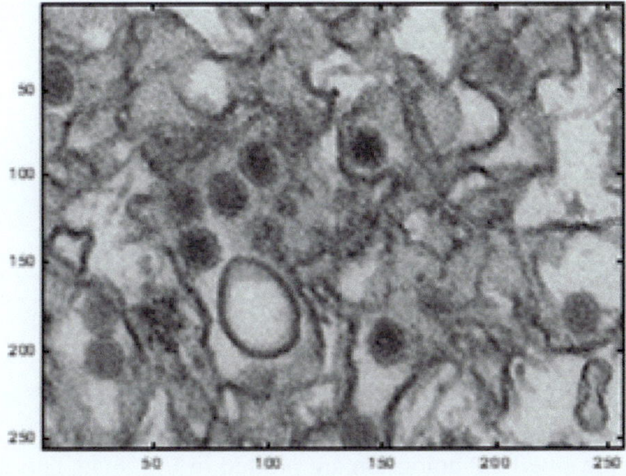

Fig. 6.1 Electron microscopy image of ZIKA virus. The particles are 40 nm in size, with an outer envelope and an inner dense core. The original image has dimensions of 652 × 434 pixels and is resized to dimensions of 256 × 256 pixels

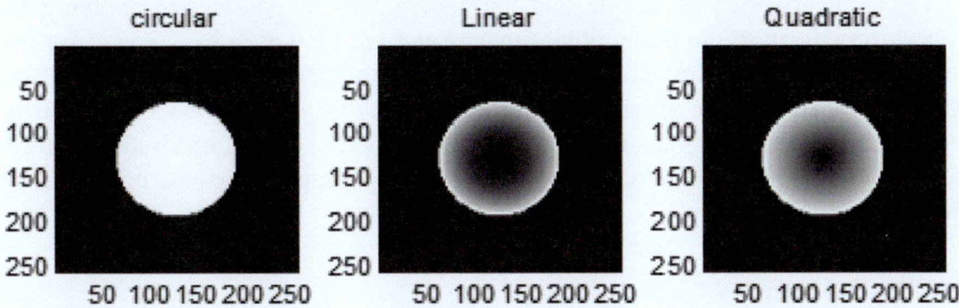

Fig. 6.2 Images of the apertures used for forming the scanning hologram

Fig. 6.3 a PSF
corresponding to the
different apertures where the
radius of 8 pixels in all
apertures is computed from
the FFT algorithm. **b** PSF
corresponds to the circular
aperture (solid line), while
the dotted curve corresponds
to the quadratic aperture.
The radial range extends
from -6π to 6π. The curves
are obtained from Eq. (6.3)

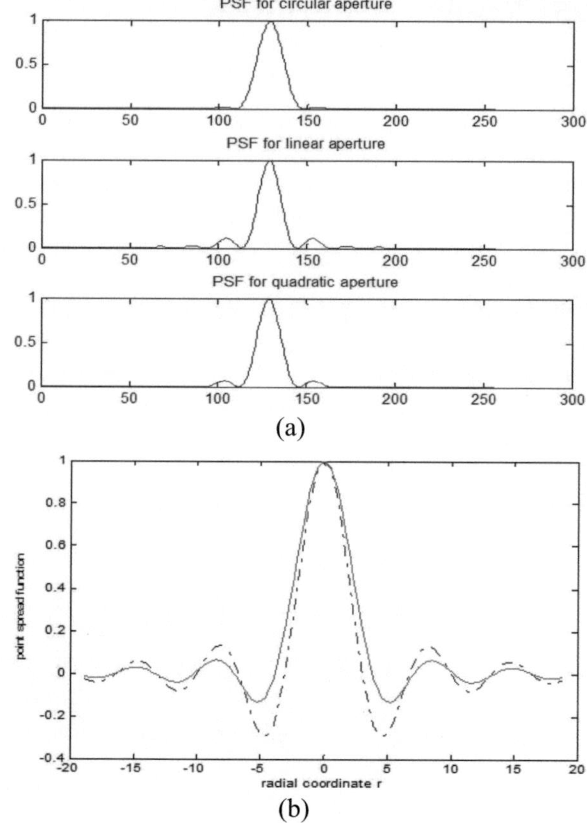

The scanning FZP holograms are plotted using Eq. (6.14). The results of the computation of the sine FZP hologram are plotted in Fig. 6.4a, and the cosine FZP hologram is plotted in Fig. 6.4b. The two pupils used in the scanning microscope shown in the preceding chapter Fig. 5.1 are the delta function and the circular uniform aperture [10]. In this chapter, the quadratic aperture replaces the constant circular aperture.

The reconstructed sine, cosine, and complex images are obtained by applying the Fourier transform to the holographic images and are plotted in Fig. 6.5a–d for the constant circular aperture, while the reconstructed images shown in Fig. 6.6a–d are plotted for the quadratic aperture. According to the PSF results shown in Fig. 6.3b, the resolution of the complex hologram image Hc^+ in the case of quadratic aperture, manipulation is better than that of the corresponding image in the case of a constant circular aperture.

Fig. 6.4 a Sine FZP
hologram using two pupils,
one with a delta function and
the other with a circular
uniform shape.
b Reconstruction of cosine
FZP holograms using two
pupils, one with a delta
function and the other with a
circular uniform shape

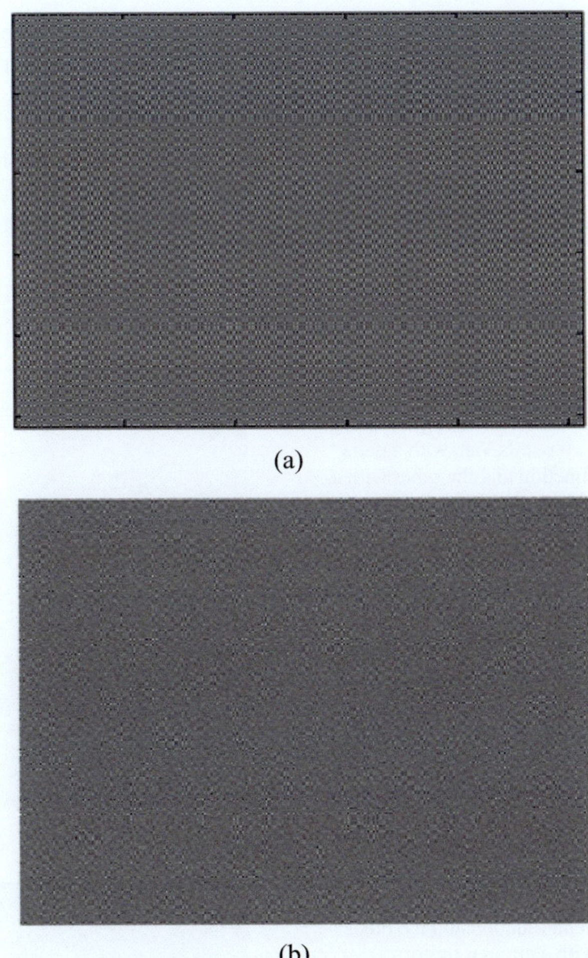

(a)

(b)

Referring to the contrast results plotted in Table 6.1, in the case of a quadratic aperture, contrast $= 0.75$ is much greater than contrast $= 0.6$ in the case of a constant circular aperture.

Hence, the reconstructed images from the complex holograms Hc^+ in the case of quadratic modulation have better contrast than the corresponding contrast obtained in the case of a constant circular aperture.

The image contrast is computed from the experimental definition of visibility as follows:

$C = \frac{(I_{\max} - I_{\min})}{(I_{\max} + I_{\min})}$. The contrast values are written in Table 6.1 for the original and reconstructed images in the case of quadratic, linear, and uniform circular apertures, in addition to the delta function as a second pupil in all cases.

A comparison between the reconstructed images in the case of the sine FZP and the complex Hc^+ holograms, as shown in Table 6.1, revealed the following results: for the sine hologram $C_{\text{circular}} = 0.53$, $C_{\text{linear}} = 0.47$, and $C_{\text{quadratic}} = 0.52$. For the complex hologram Hc^+, the contrast values are as follows: $C_{\text{circular}} = 0.604$, $C_{\text{linear}} = 0.66$, and $C_{\text{quadratic}} = 0.75$. From these results, the highest contrast for the reconstructed images corresponding to the complex FZP hologram Hc^+ is obtained from the quadratic aperture manipulation at $C_{\text{quadratic}} = 0.75$. For the sine hologram, comparable contrast is obtained in the case of a constant circular aperture and quadratic aperture. The second aperture is unchanged in all manipulations and has a delta function.

The peak signal-to-noise ratio (PSNR) is computed from the following formula:

Fig. 6.5 **a** Reconstruction of the sine FZP hologram using two pupils, one with a delta function and the other with a circular uniform shape. **b** Reconstruction of cosine FZP hologram using two pupils, one with a delta function and the other with a circular uniform shape. **c** Reconstruction of the Hc^+ complex scanning hologram of the Zika virus image using two pupils, one with a delta function and the other with a circular uniform shape. **d** Reconstruction of Hc^- complex scanning hologram of the image of the Zika virus. Two pupils, one with a delta function and the other with a circular uniform shape, are used. A severe out-of-focus image is formed at a distance $z = 2\,z_0$

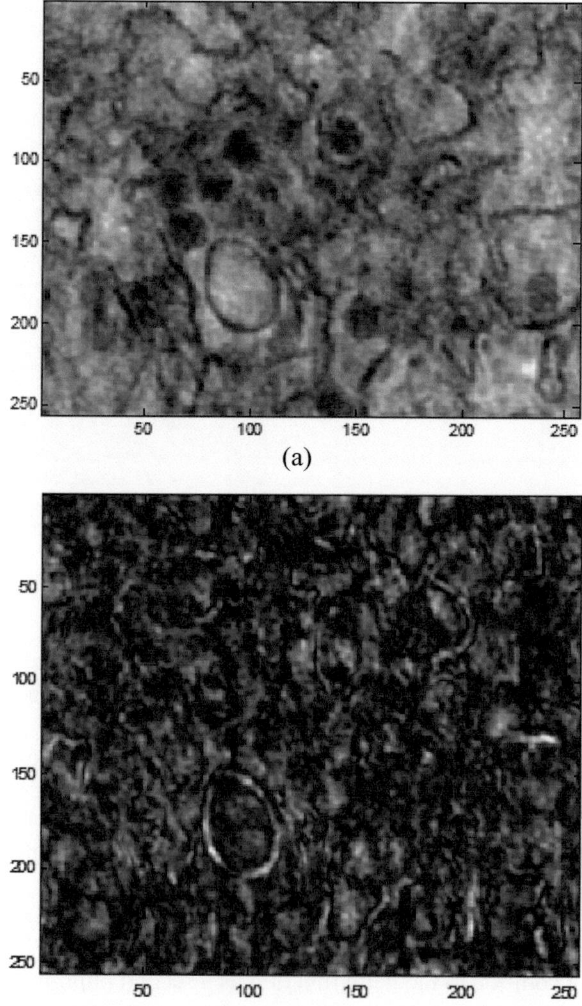

(a)

(b)

Fig. 6.5 (continued)

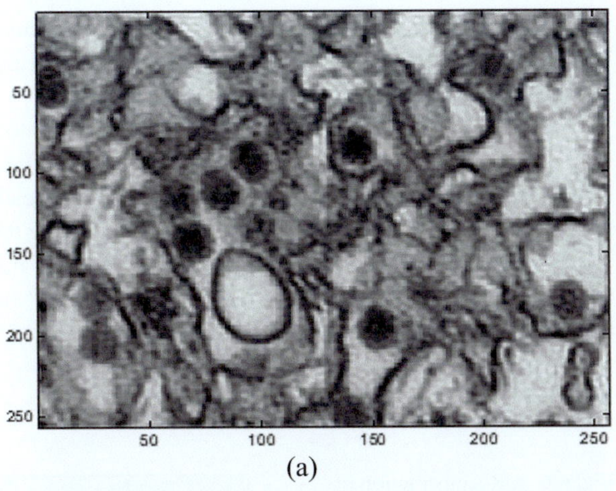

(a)

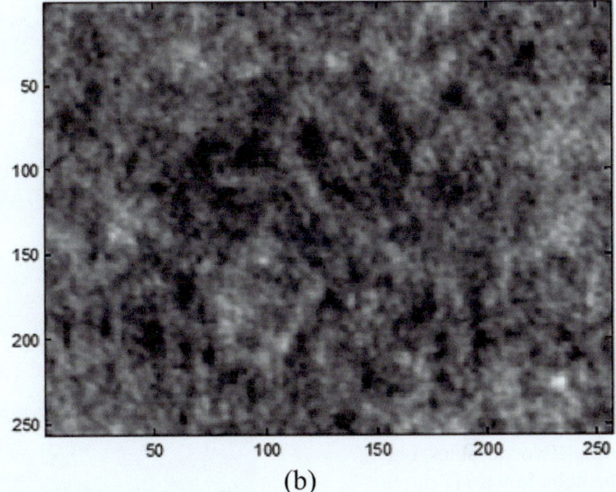

(b)

$$\text{PSNR} = 10 \log_{10}\left(\frac{R^2}{\text{MSE}}\right). \tag{6.15}$$

In the Eq. (6.15), the mean square error (MSE) is defined as:

$$\text{MSE} = \frac{\sum_{m=1}^{M} \sum_{n=1}^{N} [I_1(m, n) - I_{\text{reconst}}(m, n)]^2}{M \times N} \tag{6.16}$$

where the input image $I_1(m, n)$ and the reconstructed image from the FZP $I_{\text{reconst}}(m, n)$

R is the maximum fluctuation in the input image data type.

The PSNR results corresponding to the reconstructed FZP holograms shown in Figs. 6.5 and 6.6 are shown in Table 6.2.

Referring to the results shown in Table 6.2 and plotted in Fig. 6.7, we concluded that the optimum value of PSNR $= 26.59$ is attained in the case of the complex FZP hologram H$^+$ and using quadratic aperture instead of uniform circular aperture. The comparative results of the PSNR corresponding to the uniform circular aperture are given in Table 6.2. We obtained the following inequality for the PSNR corresponding to the reconstruction using complex FZP hologram (H^+) provided with either quadratic or uniform circular apertures:

$$\text{PSNR}_{\text{quad}} = 26.59 > \text{PSNR}_{\text{circ}} = 23.16.$$

Fig. 6.6 **a** Reconstruction of the sine FZP hologram using two pupils, one with a delta function and the other with a quadratic distribution. **b** Reconstruction of cosine FZP hologram using two pupils, one with a delta function and the other with a quadratic distribution. **c** Reconstruction of the complex FZP hologram Hc^+ using two pupils, one with a delta function and the other with a quadratic distribution. **d** Reconstruction of Hc^- complex scanning hologram of the image of the Zika virus. Two variables, one of the delta functions and the other of the quadratic distributions, are used. A severe out-of-focus image is formed at a distance $z = 2\,z_0$

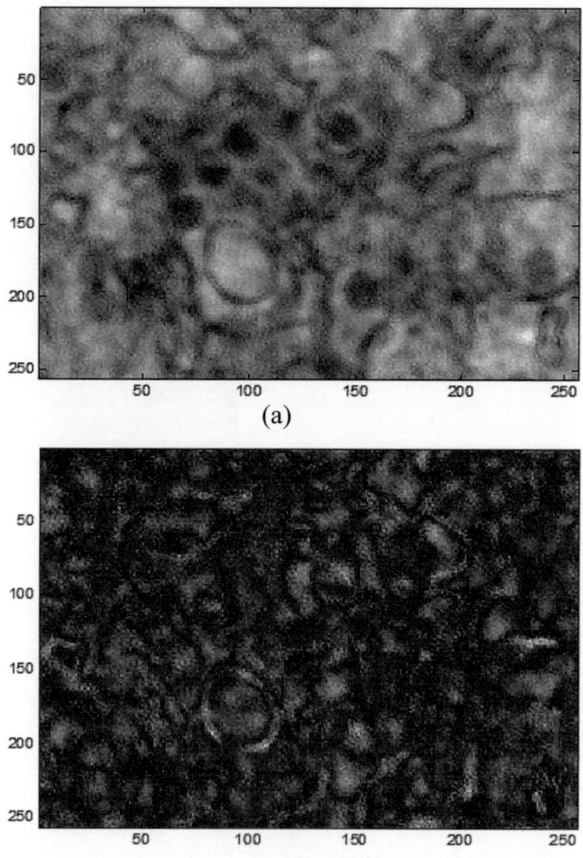

Fig. 6.6 (continued)

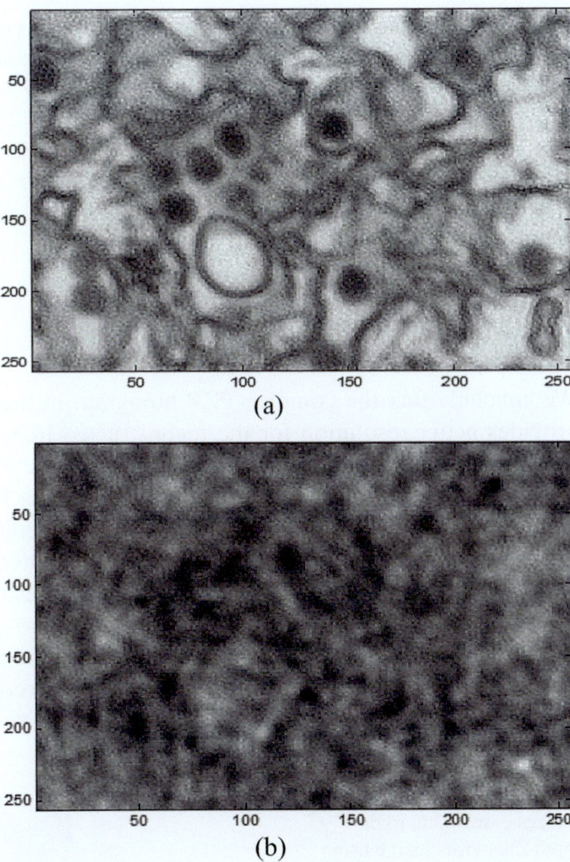

(a)

(b)

Table 6.1 Contrast values computed for the original and all the reconstruction images for the modulated linear and quadratic apertures and compared with the constant circular aperture

Contrast between the original and reconstructed images	Circular uniform aperture	Linear aperture	Quadratic aperture
Contrast C for the original image	0.81	0.81	0.81
C for the sine FZP hologram	0.53	0.47	0.52
C for the cosine FZP hologram	0.99	0.99	0.99
C for the complex hologram Hc^+	0.60	0.66	0.75
C for the complex hologram Hc^-	0.58	0.49	0.57

Table 6.2 PSNR values computed for the original Zika image and all the reconstruction images for the modulated quadratic apertures and compared with the uniform circular aperture

The PSNR is computed from the original and reconstructed images	$PSNR_{quad}$	$PSNR_{circ}$
Sine FZP hologram	19.93	21.05
Cosine FZP hologram	12.55	12.97
The complex hologram Hc^+	26.59	23.16
The complex hologram Hc^-	16.97	17.05

6.4   Conclusion

We conclude that the complex FZP hologram in the case of a quadratic aperture provides better resolution for the reconstructed images compared with the reconstructed images obtained from the sine and cosine FZP holograms in the case of a circular uniform aperture. This is attributed to the PSF improvement obtained in the case of the quadratic aperture and hence the resolution improvement. The delta function of the second pupil was the same for all investigated heterodyne scanning holograms. The contrast of the reconstructed images corresponding to the complex hologram manipulation is greater in the case of the quadratic aperture than in the case of the linear and uniform circular apertures. In addition, the peak signal-to-noise ratio (PSNR) has the greatest value in the case of the complex hologram H^+ provided with the quadratic aperture.

Fig. 6.7 a–d Reconstructed images correspond to the input Zika virus image using a quadratic aperture instead of a uniform circular aperture. In addition, the PSNR corresponds to the different real and complex holograms

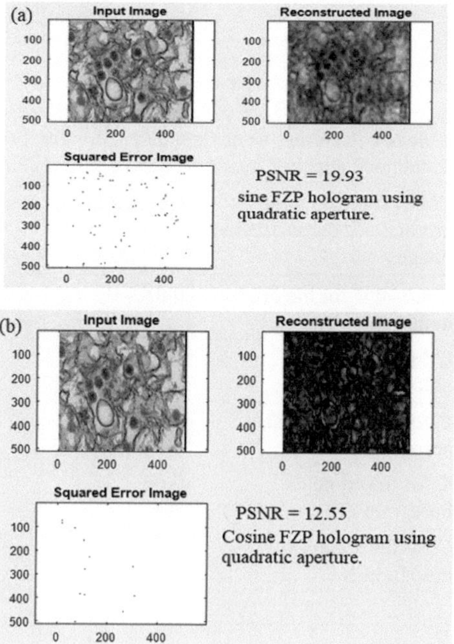

Fig. 6.7 (continued)

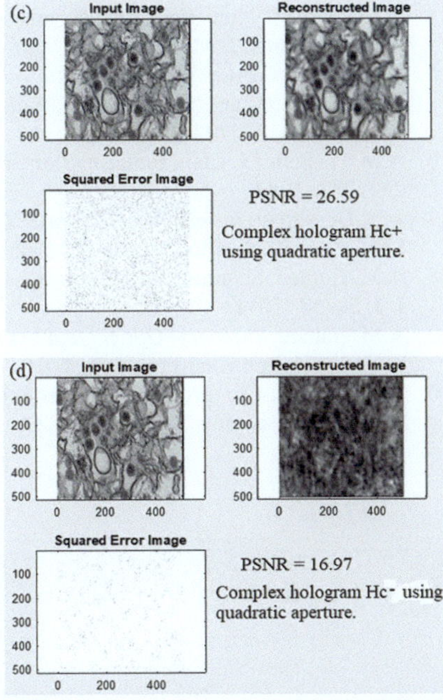

PSNR = 26.59

Complex hologram Hc+ using quadratic aperture.

PSNR = 16.97

Complex hologram Hc− using quadratic aperture.

References

1. J.W. Goodman, R.W. Lawrence, Digital image formation from electronically detected holograms. Appl. Phys. Letters **11**, 77–79 (1967)
2. A.W. Lohmann, D.P. Paris, Binary Fraunhoffer holograms, generated by computer. Appl. Opt. **6**, 1739–1748 (1967)
3. M.A. Kronrod, L. Yaroslavsky, Computer synthesis of transparency holograms. Sov. Phys. Tech. **17**, 329–332 (1972)
4. K. Nagashima, Opt. Laser Tech. **18**, 157–162 (1986)
5. A.M. Hamed, Discrimination between speckle images using diffusers modulated by some deformed apertures: simulations. Opt. Eng. **50**, 1–7 (2011). https://doi.org/10.1117/1.3530085
6. B.B. Gorbatenko, L.A. Maksimova, V.P. Ryabukho, Opt. Spectrosc.Spectrosc. **106**, 281–287 (2009)
7. N. Takanori, O. Mitsukiyo et al., Image quality improvement of digital holography by superposition of reconstructed images obtained by multiple wavelengths. Appl. Opt. **47**, D38–D43 (2008)
8. J.P. Liu, T.C. Poon, Two-step-only quadrature phase-shifting digital holography. Opt. Lett. **34**, 250–252 (2009)
9. T.C. Poon, Scanning holography and two-dimensional image processing by acoustic-optic two-pupil synthesis. J. Opt. Soc. Am. A **4**, 521–527 (1985)
10. T.C. Poon, *Optical Scanning Holography with MATLAB* (Springer Science& Business Media, 2007)
11. T.C. Poon, J. Holography Speckle **1**, 6–25 (2004)
12. A.W. Lohmann, T. Rhodes, Two-pupil synthesis of optical transfer functions. Appl. Opt. **17**, 1141–1150 (1978)
13. T.C. Poon, T. Kim et al., Twin-image elimination experiments for three-dimensional images in optical scanning holography. Opt. Lett. **25**, 215–217 (2000)

14. A.M. Hamed, Numerical speckle images formed by diffusers using modulated conical and linear apertures. J. Mod. Opt.Mod. Opt. **56**, 1174–1181 (2009)
15. A.M. Hamed, Formation of speckle images formed for diffusers illuminated by modulated apertures (circular obstruction). J. Mod. Opt. **56**, 1633–1642 (2009). https://doi.org/10.1080/09500340903277792
16. A.M. Hamed, J.J. Clair, Image and super-resolution in optical coherent microscopes. Optik **64**, 277–284 (1983)
17. J. W. Goodman, *Introduction to Fourier Optics and Holography*, 3rd ed. (Roberts and Company Publishers, Englewood, Colorado, 2005)
18. A.M. Hamed, Scanning holography using a modulated linear pupil: simulations. Opt. Photon. J. **1**, 52–58 (2011)
19. A.M. Hamed, *Topics on Optical and Digital Image Processing Using Holography and Speckle Techniques* (Publisher Lulu.com, 2015). ISBN: 9781329328464
20. A.M. Hamed, Compromising of resolution and contrast using quadratic aperture in scanning holographic imaging. Int. J. Photon. Opt. Tech. (IJPOT) **2**, 18–23 (2016)

Chapter 7
Processing of Cancerous Mammographic Images Using Improved Fourier Holograms

In this chapter, improved reconstruction from the coded Fourier hologram of cancerous mammographic images using the Wiener filter is obtained. In addition, image segmentation of cancerous images using a median filter, removal of small objects, and removal of connected objects on the border of the image are investigated using multiple-beam interference.

7.1 Introduction

An improved reconstruction from the Fourier hologram using a spatial light modulator (SLM) was investigated. The method is based on applying digital prefiltering of the encoded distribution of the SLM to compensate for the signal distortion caused by the finite pixel size of the SLM [1]. Recently, the processing of plasma images using Fourier holograms was considered in [2–12], where improved reconstructed images were obtained using Wiener filtering.

In recent decades, many applications based on measuring the fringe shift in optical and synthetic fibers to obtain refractive index information [13–26] have been outlined. Recently, digital two and multiple-beam interference methods have been applied to coronavirus images and other medical images [27, 28] to extract the refractive index, which is related to fringe shifts. In addition, a cascaded two-beam interference arrangement is used to determine the intensity of higher orders of $\cos^2 \theta$ in the form of $\cos^{2n} \theta$, where n represents the number of feedback passes to the interferometer [29, 30].

In this chapter, an improved reconstruction of mammographic images considering Fourier holograms is obtained by filtration, as in [12]. The segmented cancerous zones obtained from the reconstructed images are investigated. In addition, image processing of cancerous mammographic images with interferometric images corresponding to the segmented cancer was investigated.

© The Author(s), under exclusive license to Springer Nature Switzerland AG 2025

A. Hamed, *Holographic Imaging Using Aperture Modulation*,

SpringerBriefs in Applied Sciences and Technology,

https://doi.org/10.1007/978-3-031-96989-8_7

7.2 Theoretical Analysis

The mammographic image is recorded numerically as a Fourier hologram using MATLAB code as follows: a diffuser of the same dimensions as the rescaled grayscale image of dimensions 512×512 pixels is constructed numerically. Then, the multiplication of both the cancerous mammographic image and the diffuser was Fourier transformed numerically to obtain the holographic interference image. It is represented by the convolution product of the Fourier transform of the mammographic image and the speckle pattern. The speckle pattern is computed by performing the Fourier transform on the diffuser. Third, the inverse Fourier transformation is applied to the described Fourier hologram to obtain the reconstructed cancerous mammographic images.

The two Fourier transform operations are summarized as follows:

First Step: Recording the Amplitude and Phase of the Hologram.

The complex amplitude of the image is analytically represented as:

$$A(x, y) = a(x, y) \exp\left[j\Phi(x, y)\right]. \tag{7.1}$$

It is numerically written as follows:

$$A(x, y;\ \lambda) = \sum_{n=1}^{N} \sum_{m=1}^{M} a(m\Delta x, n\Delta y) \exp\left[j\Phi(m\Delta x, n\Delta y;\ \lambda_i)\right], \tag{7.2}$$

where the continuous variables (x, y) are replaced by numerical values as follows: $x = m\,\Delta x$ and $y = n\Delta y$; $j = \sqrt{-1}$.

Similarly, the diffuser is numerically written as follows:

$$D(x, y;\ \lambda) = d(m\Delta x, n\Delta y) \exp\left[(j)\mathrm{rand}\,(m\Delta x, n\Delta y;\ \lambda_i)\right], \tag{7.3}$$

where rnd is a random value that ranges from 0 to 1, and $d(m\Delta x, n\Delta y)$ is the amplitude weighting factor. For example, if the image has square dimensions of 2 cm in height and 2 cm in width and the diffuser has the same dimensions, then $\Delta x = \Delta y = 18$ μm.

The complex amplitude of the Fourier hologram is obtained by operating the FFT upon the multiplication product of the two matrices represented in Eq. (4.2) and Eq. (4.3).

Hence, the complex amplitude of the hologram $B(u, v)$ is represented as follows:

$$B(u, v) = \mathrm{F.T.}\left[A(x, y;\ \lambda)D(x, y;\ \lambda)\right]$$

$$= \mathrm{F.T.}[A(x, y;\ \lambda)] \otimes \mathrm{F.T.}[D(x, y;\ \lambda)] \xrightarrow{\text{yields}} A_{\mathrm{holo}}(u, v;\ \lambda) \otimes D_{\mathrm{holo}}(u, v;\ \lambda), \tag{7.4}$$

where (u, v) are the spatial coordinates in the Fourier plane. $\otimes$ is a symbol for convolution.

Second Step: Reconstruction Process (Inverse Fourier Transform)

$$c(x', y') = \text{F.T}^{-1}[A_{\text{holo}}(u, v) \otimes D_{\text{holo}}(u, v)] = A_{\text{reconst}}(x', y') \cdot D_{\text{reconst}}(x', y'), \tag{7.5}$$

where (x', y') represent the Cartesian coordinates in the imaging plane or the reconstruction plane. The numerical reconstruction images have maximum dimensions at $x' = x'_{\max} = M\, \Delta x'$ and $y' = y'_{\max} = N\, \Delta y'$ for the matrix of dimensions of 1024 $\times$ 1024 pixels.

7.2.1 The Isolation of the Cancer from the Mammographic Image Using Segmentation

The following steps are considered for the segmentation process:

1. Preprocessing was performed using the median filter.
2. Determine the connected components.
3. Compute the area of each component.
4. Remove small objects.
5. The connected objects on the border are removed.

The cell of interest has been successfully segmented, but it is not the only object that has been found. Any objects that are connected to the border of the image can be removed using the imclearborder function.

6- Interference of the segmented image.

All these operations are executed using MATLAB codes.

7.2.2 Segmented Images Modulated by Multiple Beam Interference

The intensity distribution in the case of an ordinary FPI is given by the following formula [23]:

$$I(\delta; R) = \frac{T^4}{1 + R^4 - 2R^2 \cos(\delta)}, \tag{7.6}$$

where T is the transmission coefficient and R is the reflection coefficient of the interferometer. δ is the phase difference between any two adjacent emerging rays.

Equation (7.6) is used in the modulation of the segmented images.

7.3   Results and Discussion

The input cancerous mammographic image of dimensions 512 × 512 pixels used in the formation of the coded image is shown in Fig. 7.1. The Fourier hologram considered for coding the image is fabricated using Eq. (7.4) and the MATLAB code. The hologram is plotted in Fig. 7.2. It has dimensions of 1024 × 1024 pixels. The reconstructed conjugate images obtained from the hologram before filtration using Eq. (7.5) are shown in Fig. 7.3a. Equation (7.5) shows that the noise in the reconstructed image is due to the superimposition of the diffuser over the image. The same plots of improved image contrast where we eliminate noise using the Wiener filter are shown in Fig. 7.3b.

The input image used for cancer isolation in the segmentation process is shown in Fig. 7.1. The preprocessed image using the median filter is shown in Fig. 7.4a.

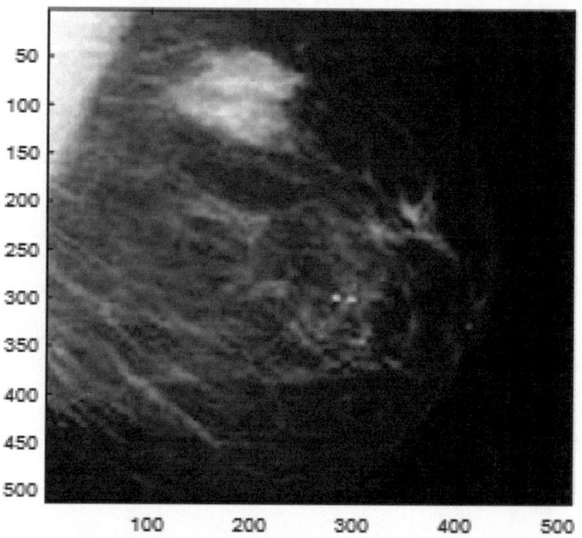

Fig. 7.1 Image of the cancerous mammographic image with dimensions 512 × 512 pixels

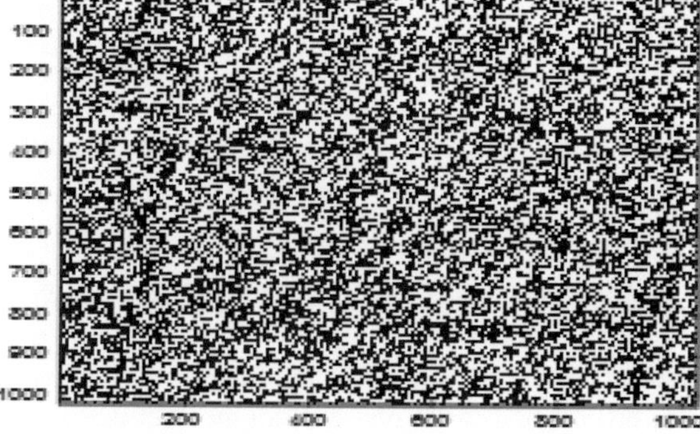

Fig. 7.2 Fourier hologram of the mammographic image shown in Fig. 7.1. The hologram of the coded image has dimensions of 1024 × 1024 pixels

Fig. 7.3 **a** Reconstructed
conjugated images
corresponding to the
cancerous mammographic
image before filtration. The
images are affected by
multiplicative noise.
b Reconstructed conjugated
images corresponding to the
cancerous mammographic
image after filtration. A
Wiener filter was used for
image filtration

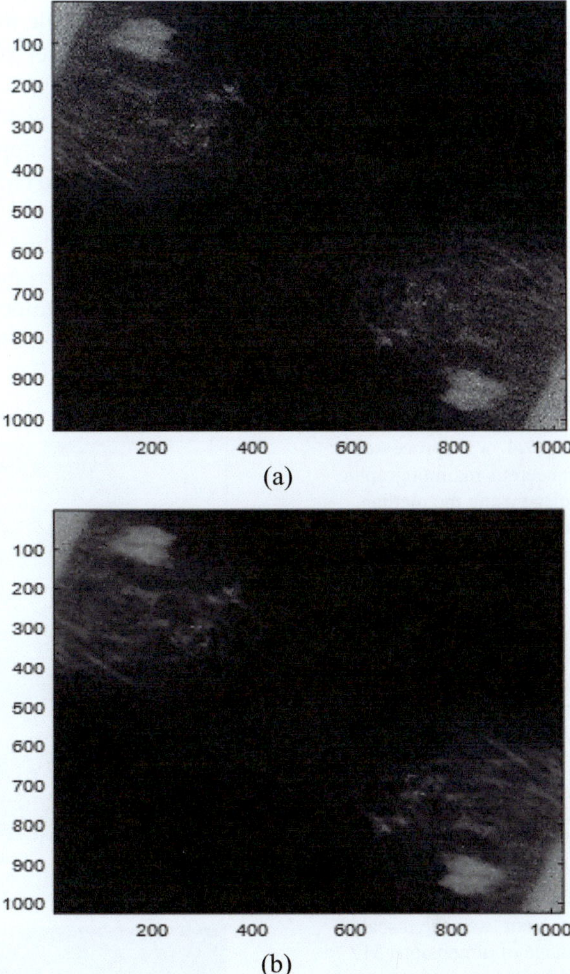

The determination of the connected components in the preprocessed image is given
in Fig. 7.4b. The removal of small objects from the connected components in the
preprocessed image is plotted in Fig. 7.4c. The removal of connected objects on
the border is shown in Fig. 7.4d. The rescaled image after the removal of connected
objects on the border of dimensions 512×512 pixels is given in Fig. 7.4e. Figure 7.4f:
The interference fringes modulated by the black and white (B/W) image are plotted
as in Fig. 7.4f, while those modulated by the mammographic segment are shown
in Fig. 7.4g. A binary image of segmented cancer in the mammographic image is
plotted in Fig. 7.4e, while a grayscale image of the cancer is shown in Fig. 7.5. The
noise appeared to the right of the image far from the cancer segment.

The line plot of the cancerous segment shown in Fig. 6.4g from the rescaled
image of dimensions 512×512 pixels is given in Fig. 7.6. The horizontal range of

the segment extends from 103 to 235 pixels, while the vertical range extends from 37 to 137 pixels. The vertical sections of the three plots at 60, 85, and 110 pixels are shown in Fig. 7.6. The different locations shown in the plots provide information about the topography of the cancer segment as well as its shape. For example, a midpoint located at 165 pixels for two close peaks appeared in the vertical line at 60 pixels. However, it is located at 173 pixels for two close peaks corresponding to the vertical line at 85 pixels.

A similar procedure for segmentation is applied to the tumor mammographic image, as shown in Fig. 7.7. The results of interferometric segmentation are plotted in Fig. 7.8a–g. A binary image of the segmented tumor is plotted in Fig. 7.8e, while a grayscale image of the tumor is shown in Fig. 7.9. The line plots for the segmented tumor in the mammographic image are shown in Fig. 7.10a, b.

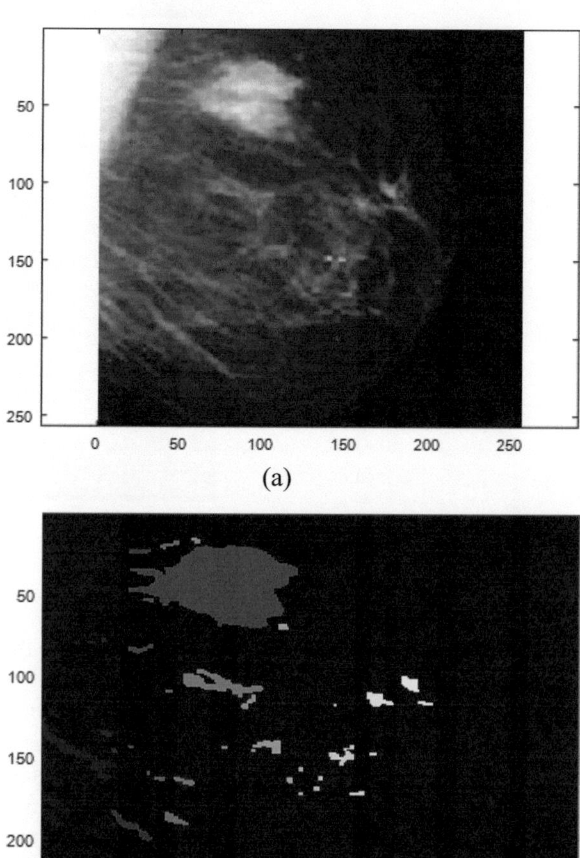

Fig. 7.4 **a** Preprocessed cancerous mammographic image using the median filter. **b** Determination of the connected components in the preprocessed image. **c** Removal of small objects from the connected components in the preprocessed image. **d** Removal of connected objects on the border. **e** Rescaled image after the removal of connected objects on the border of 512 × 512 pixels. **f** Interference fringes modulated by the black and white (B/W) cancerous segment from the rescaled image of dimensions 512 × 512 pixels. **g** Interference fringes modulated by the cancerous segment from the image

Fig. 7.4   (continued)

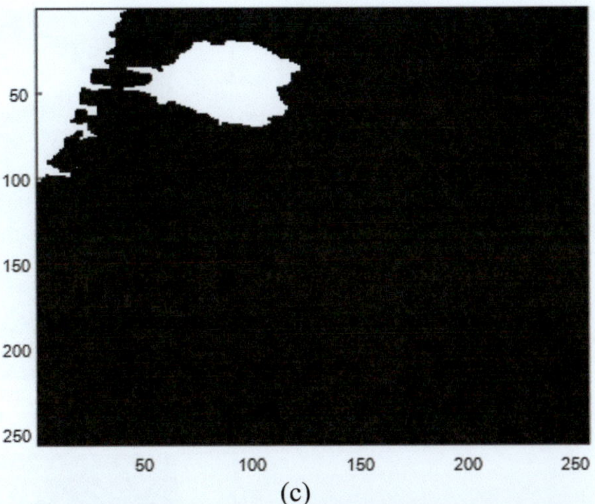

(c)

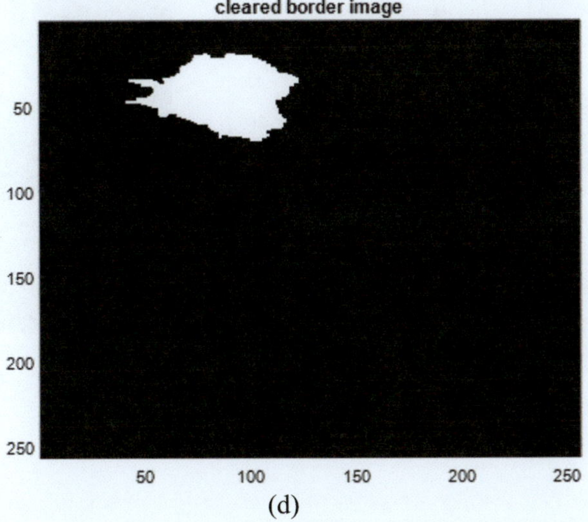

(d)

Fig. 7.4 (continued)

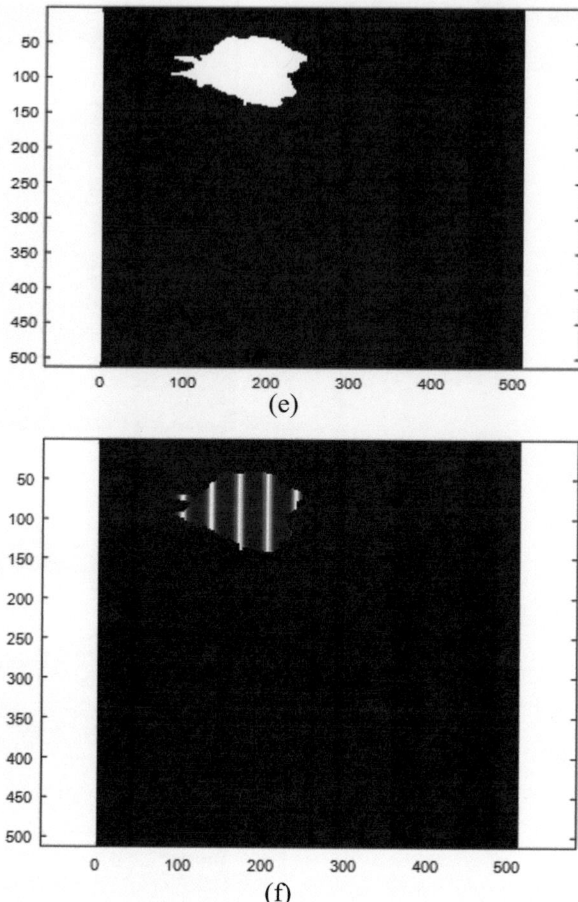

(e)

(f)

Fig. 7.4   (continued)

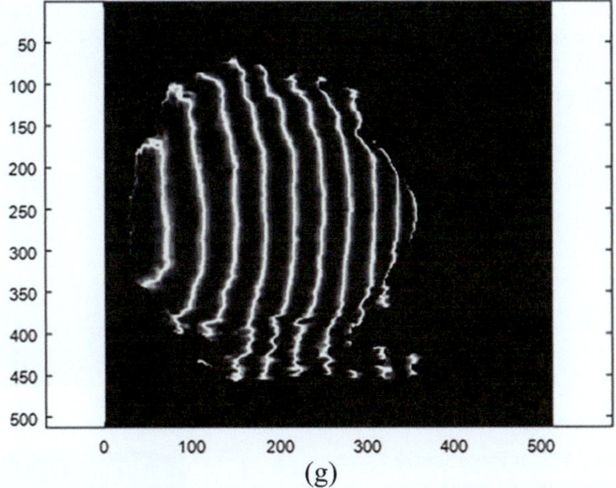

Fig. 7.5   Grayscale image of the isolated segmented cancer in the mammographic image. Noise appears from the computations to the right of the image

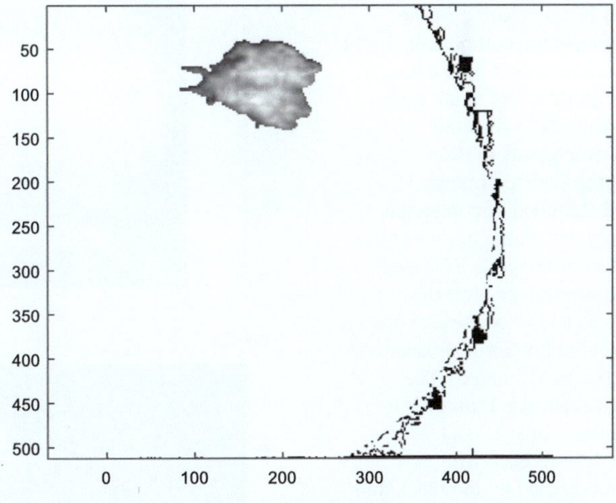

Fig. 7.6   Line plot of the cancerous segment from the interferometric image with dimensions 512×512 pixels. The horizontal range of the segment extends from 103 to 235 pixels, while the vertical range extends from 37 to 137 pixels

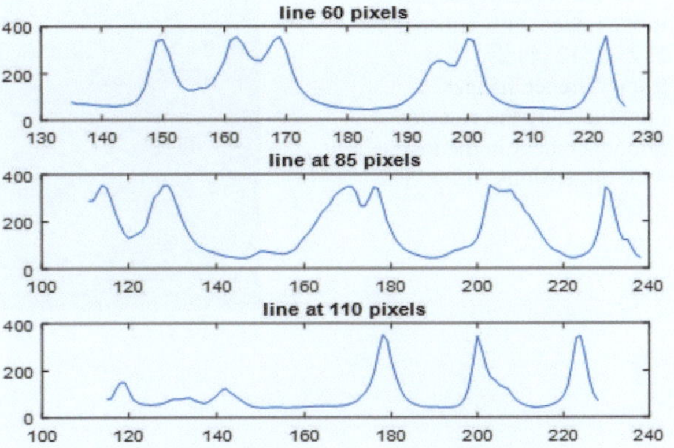

Fig. 7.7 Tumor in the mammographic input image with dimensions of 256 × 256 pixels

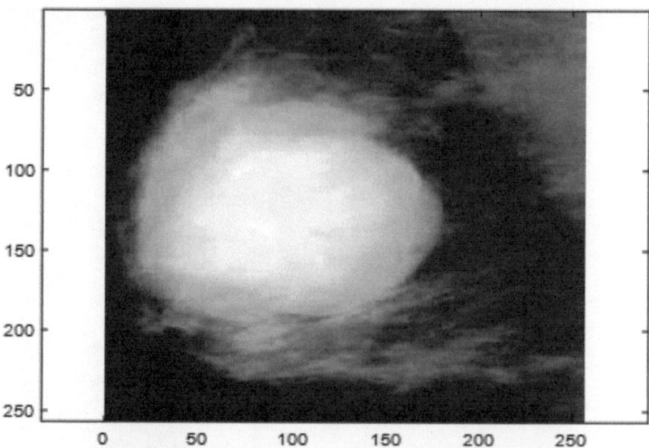

Fig. 7.8 **a** Preprocessed tumor image using the median filter. **b** Determination of the connected components in the preprocessed tumor image. **c** Removal of small objects from the connected components in the preprocessed image. **d** Removal of connected objects on the border of the tumor image. **e** Rescaled tumor image after the removal of connected objects on the border of dimensions 512 × 512 pixels. The reconstructed tumor is a binary image [0,1]. **f** Interference fringes modulated by the black and white (B/W) tumor segment in the image with dimensions 512 × 512 pixels. **g** Interference fringes modulated by the rescaled tumor segment in the image with dimensions 512 × 512 pixels.

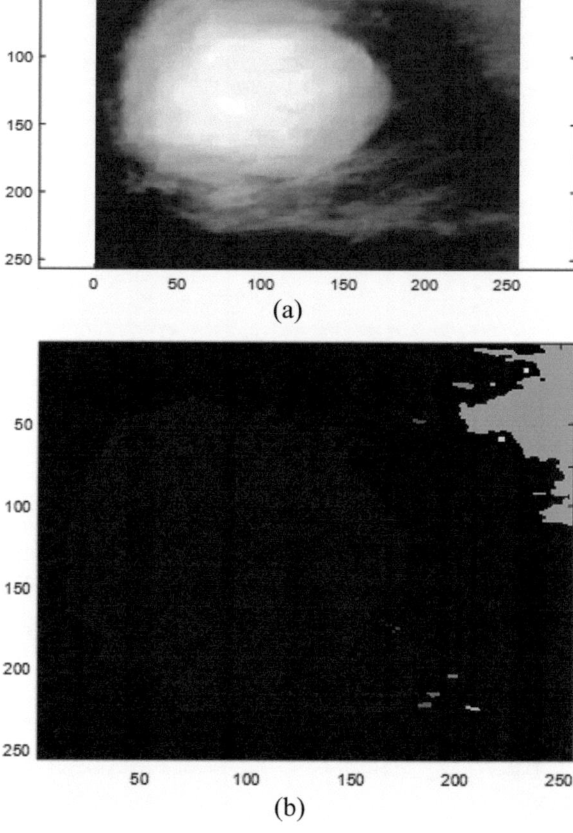

(a)

(b)

Fig. 7.8 (continued)

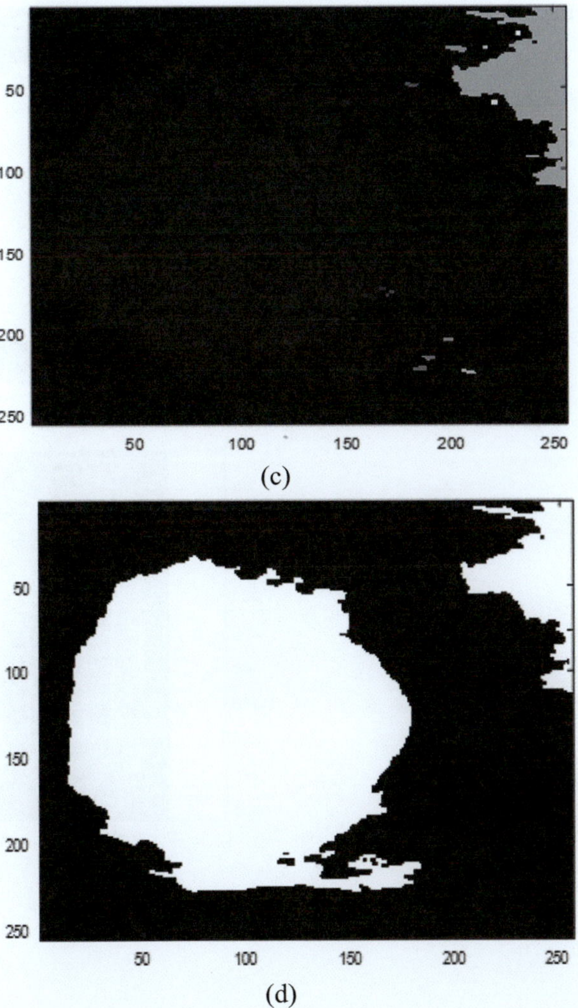

Fig. 7.8 (continued)

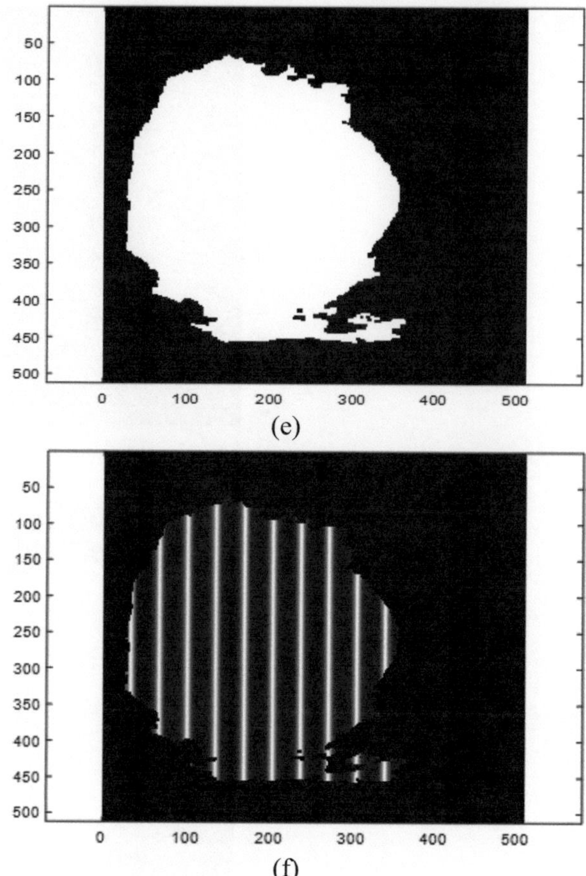

(e)

(f)

Fig. 7.8 (continued)

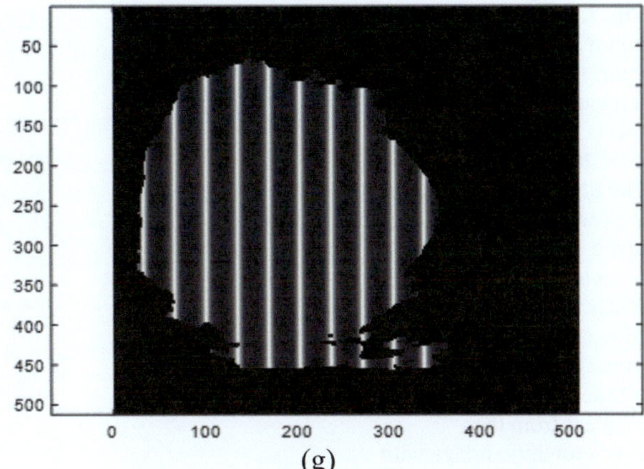

(g)

Fig. 7.9 Reconstructed tumor in grayscale image after segmentation

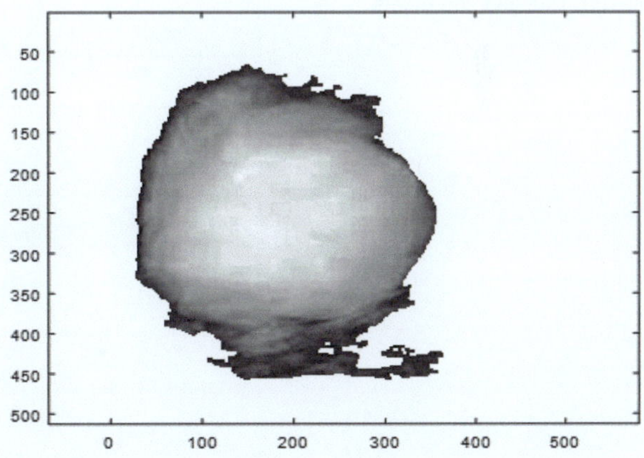

Vertical sections of the three plots at 200, 250, and 300 pixels are shown in Fig. 7.10a. Another three plots corresponding to the tumor segment at 100, 150, and 200 pixels, as shown in Fig. 7.10b, are much shifted, showing that the fringe shift occurred from the tumor image at different locations. For example, only 4 peaks are shown at 100 pixels, while 7 and 9 peaks are located at 150 and 200 pixels, respectively. Hence, the maximum tumor width corresponds to the maximum number of peaks, as shown in Fig. 7.10a, for the plots at 200, 250, and 300 pixels.

Fig. 7.10 a Line plot of the tumor segment from the interferometric image with dimensions 512×512 pixels. The tumor was found near the center of the image. The plot at the tumor center at nearly 250 pixels is shown in the middle, while the other two plots at 200 and 300 pixels are shifted due to the fringe shift that occurred from the tumor image. **b** Line plot of the tumor segment from the interferometric image with dimensions 512×512 pixels. Three plots are shown at 100, 150, and 200 pixels

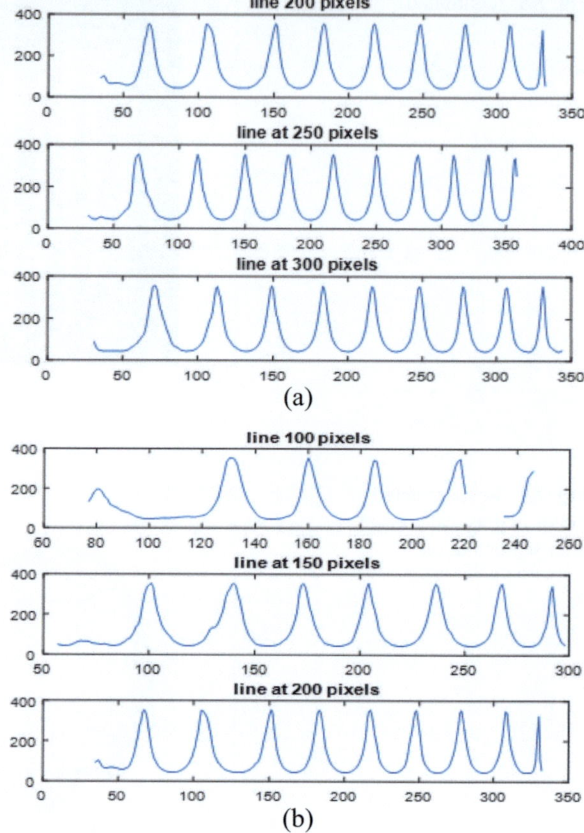

7.4 Conclusion

First, we fabricated coded cancerous and tumor mammographic images, without segmentation, using improved Fourier holography.

Second, we apply the segmentation procedure and multiple-beam interference to the cancerous and tumor images to obtain information about the cancer and tumor shapes from the interferometric segments.

Finally, the tumor and cancer segments are obtained as grayscale images and interferometric images.

References

1. M. Agour, C. Falldorf, C. von Kopylov, Digital prefiltering approach to improve optical reconstructed wave fields in opt-electronic holography. J. Opt. **12**, 1–7 (2010)
2. J.W. Goodman, *Introduction to Fourier Optics and Holography*, 3rd edn. (Roberts & Company Publishers, Greenwood Village, 2005)
3. A.M. Hamed, Polychromatic image processing using thick holographic multiplexed filter. Optica Applicate, **13**, 205–213 (1983)
4. R.N. Bracewell, Fourier Transform and Its Applications (McGraw-Hill, New York, 1978)
5. A.E. Macgregor, Computer generated holograms from dot matrix and laser printers. Am. J. Phys. **60**, 839–846 (1992). https://doi.org/10.1119/1.17067
6. S. Trester, Computer simulated holography and computer-generated holograms. Am. J. Phys. **64**, 472–476 (1996). https://doi.org/10.1119/1.18194
7. G. Pedrini, S. Schedin, H.J. Tiziani, Lensless digital holographic interferometry for measurement of large objects. Opt. Commun. **171**, 29–36 (1999). https://doi.org/10.1016/S0030-4018(99)00486-1
8. C. Wagner, S. Seebacher, W. Osten, W. Jüptner, Digital Recording and numerical reconstruction of lensless fourier holograms in optical metrology. Appl. Opt. **38**, 4812–4820 (1999). https://doi.org/10.1364/AO.38.004812
9. A.C. Bovik, S.T. Acton, Basic Linear Filtering with application to image enhancement, in *Handbook of Image and Video Processing*, ed. by A.C. Bovik (Academic, San Diego, 2000)
10. M. Agour, E. Kolenovic, C. Falldorf, C. von Kopylov, Suppression of higher diffraction orders and intensity improvement of optically reconstructed holograms from a spatial light modulator. J. Opt. A Pure Appl. Opt. **11**, Article ID: 105405 (2009). https://doi.org/10.1088/1464-4258/11/10/105405
11. A.M. Hamed, Scanning holography using a modulated linear pupil: simulation. Opt. Photon. J. **1**, 52–58 (2011). https://doi.org/10.4236/opj.2011.12008
12. A.M. Hamed, Holographic imaging of Argon plasma images. Opt. Photon. J. **4**, 136–142 (2014). https://doi.org/10.4236/opj.2014.46014
13. N. Barakat, S. Mokhtar, On the factors contributing to the formation of multiple beam Fizeau fringes. J. Opt. Soc. Am. **53**, 300–301 (1963)
14. N. Barakat, A.M. Hindelah, Determination of the refractive indices, birefringence, and tensile properties of normal viscose rayon fibers. Textile Res. J. **41**, 581–584 (1964)
15. N. Barakat, Interferometric studies on Fibers: Part I: Theory of interferometric determination of Indices of Fibers. Textile Res. J. **41**, 167 (1971)
16. M.M. El Nicklawy, I.M. Fouda, An analysis of Fizeau fringes crossing a fiber with multiple skins. Textile Res. J. **71**, 252–256 (1980)
17. A. Hamza, T.Z.N. Sokkar, M.A. Abeel, Interferometric determination of optical properties of fibers of irregular transverse section and having a skin-core structure. J. Phys. D **18**, 1773 (1985)
18. A. Hamza, T.Z.N. Sokkar, M.A. Abeel, Interferometric determination of refractive indices and birefringence of fibers with irregular transverse sections. J. Phys. D **19**, 119 (1986)
19. N. Barakat, A. Hamza, A. Goned, Multiple beam interference fringes applied to GRIN optical waveguides to determine fiber characteristics. Appl. Opt. **24**, 4383–4386 (1985)
20. L.M. Boggs, H.M. Presby, D. Marcuse, Rapid automatic index profiling of whole fiber samples: part I. Bell Syst. Tech. J. **58**, 867–882 (1979)
21. H.M. Presby, D. Marcuse, H.W. Astle, L.M. Boggs, Rapid automatic index profiling of whole fiber samples: part II. Bell Sys. Tech. J. **58**, 883–902 (1979)
22. A.M. Hamed, Fourier imaging of unclad fibers using a liquid wedge interferometer. Opt. Applicate **27**, 229–240 (1997)
23. A.M. Hamed, Modeling of the fringe shift in multiple beam interference for glass fibers, Pramana. J. Phys. **70**, 643–648 (2008)
24. N. Barakat, A. Hamza, *Interferometry of Fibrous Materials* (Adam Hilger, Ltd., Techno House, Bristol, 1990)

25. M. Born and E. Wolf, *Principles of Optics*, 2nd ed. (Pergamon Press Ltd., Oxford, 1964), p 328
26. A.M. Hamed, *Some applications of scattered light and multiple beam interference using coherent light*, M.Sc. Thesis (1976)
27. A. M Hamed, Investigation of SIDA virus (HIV) images using interferometry and speckle techniques. Int. J. Innovative Res. Comp. Sci. Tech. **4**, 38–45 (2016)
28. A.M. Hamed, Image processing of coronavirus using interferometry. Opt. Photon. J. **6**, 75–86 (2016)
29. A.M. Hamed, Topics on optical and digital image processing using holography and speckle techniques, published by Lulu.com, ISBN 9781329328464, 29 Nov 2015
30. A.M. Hamed, A modified Michelson interferometer and an application on microscopic imaging. Int. J. Photon. Opt. Technol. **3**, 67–71 (2017)

Chapter 8
Fourier Holographic Imaging of Modulated Apertures

We investigate the holographic images formed using different apertures and a fixed diffuser. We computed the point spread function (PSF) for the Straubel aperture and summarized the PSF results for the linear, quadratic, black and white (B/W), hexagonal, Cauchy, and Straubel apertures. We considered these apertures in the formation of the Fourier holograms.

We computed the Fourier holograms corresponding to the modulated apertures. These apertures are linear, quadratic, black and white, Cauchy, and Straubel distributions. We presented the point spread function corresponding to these apertures as an indication of resolution. We compared the PSF results to those corresponding to the circular aperture. We discussed the contrast of the reconstructed images for the different diameters and different apertures. We showed that for a certain diameter of the aperture, the contrast of the reconstructed images is increased at the expense of the resolution. In addition, we present the reconstructed image of the Straubel filter as an aperture from the phase hologram using iteration.

8.1   Introduction

Digital holography is a new technique with various applications in physics [1], technology [2–5], biology [6], and arts [7] as well. It has the advantage of software compared to the ordinary hologram formed by hardware components. A more practical method for creating Fourier holograms is the so-called lensless configuration [8]. Recently, a Fourier-transformed digital holography is presented in [9]. It is a hybrid arrangement, half digital half analog. The Fourier hologram was constructed using analog means. The hologram was recorded digitally by a CCD camera instead of the recording medium. The reconstruction of the recorded hologram is digitally realized using the discrete Fourier transform.

© The Author(s), under exclusive license to Springer Nature Switzerland AG 2025
A. Hamed, *Holographic Imaging Using Aperture Modulation*,
SpringerBriefs in Applied Sciences and Technology,
https://doi.org/10.1007/978-3-031-96989-8_8

In this chapter, we investigate the Fourier holograms for linear, quadratic, and Straubel apertures. We reconstruct the images of the apertures and compute the corresponding contrast.

8.2 Theoretical Analysis

8.2.1 PSF for Some Modulated Apertures

For the Straubel aperture, we compute the PSF as follows:

The other modulated apertures and the corresponding PSF are summarized in Table 8.1.

Table 8.1 PSF formulae corresponding to some modulated apertures compared to those corresponding to the transparent circular aperture

The aperture	PSF formula
1. Circular $P(\rho) = 1; \ \left\lvert\frac{\rho}{\rho_0}\right\rvert \leq 1$	$h(w) = \frac{2J_1(w)}{w}$
2. Annular $P(\rho) = P_{\text{ext.}}(\rho) - P_{\text{int.}}(\rho); \ \Delta\rho = \rho_{\text{ext.}} - \rho_{\text{int.}}$	$h(w) = 2\left[\frac{J_1(w)}{w} - \varepsilon^2\frac{J_1(\varepsilon w)}{\varepsilon w}\right]; \ \varepsilon = \frac{\rho_{\text{int.}}}{\rho_{\text{ext.}}}$
3. Linear $P(\rho) = \rho; \ \left\lvert\frac{\rho}{\rho_0}\right\rvert \leq 1$	$h(w) = \frac{J_1(w)}{w} + \frac{J_0(w)}{w^2} - 2\frac{\sum_i J_i(w)}{w^2}$
4. Quadratic $P(\rho) = \rho^2; \ \frac{\rho}{\rho_0} \leq 1$	$h(w) = \frac{J_1(w)}{w} - 2\frac{J_2(w)}{w^2}$
5. Straubel $P(\rho) = 1 - \beta\left(\frac{\rho}{\rho_0}\right)^2; \ \left\lvert\frac{\rho}{\rho_0}\right\rvert \leq 1$	$h(w) = \frac{J_1(w)}{w} + 3\frac{J_2(w)}{w^2}; \ \beta = 0.6$
6. Cauchy $P(\rho) = \frac{1}{1+\beta^2\left(\frac{\rho}{\rho_0}\right)^2}; \ \left\lvert\frac{\rho}{\rho_0}\right\rvert \leq 1, \beta < 1$	$h(w) = \left(\frac{\pi}{\lambda f \beta}\right)\exp\left[-\frac{2\pi\rho_0}{\lambda f \beta}\lvert r \rvert\right]; \ w = \frac{2\pi\rho_0 r}{\lambda f}$
7. Hexagonal $P(x, y) = \text{rect}(x, y) + \text{tri}(x, y - 2a/3) + \text{tri}(x, y - 2a/3)$	$h(u, v) = \left[\frac{\sin(\pi bu)}{\pi bu}\right]\left[\frac{\sin(\pi av)}{\pi av}\right]\left\{1 + 2\left[\frac{\sin(\pi bu)}{\pi bu}\right]\left[\frac{\sin(\pi av)}{\pi av}\right]\cos\left(\frac{4\pi av}{3\lambda f}\right)\right\}$
8. Linear–conic (triangular function)	$h(w) = 2\pi\left\{\frac{J_1(0.5w)}{(0.5w)} + \frac{J_0(0.5w)}{(0.5w)^2} - 2\frac{\sum_i J_i(o.5w)}{(0.5w)^2}\right\} +$ $2\pi\left\{\frac{[2\sum_i J_i(w) - wJ_0(w)]}{w^3}\right\} - 2\pi\left\{\frac{[2\sum_i J_i(0.5w) - 0.5wJ_0(0.5w)]}{(0.5w)^3}\right\}$

We represent the Straubel filter as follows:

$$P(\rho) = 1 - \beta\left(\frac{\rho}{\rho_0}\right)^2; \quad \left|\frac{\rho}{\rho_0}\right| \leq 1. \tag{8.1}$$

ρ is the radial coordinate in the aperture plane (u, v), ρ_0 is the aperture radius, and β is a parameter.

We compute the point spread function by operating the Fourier transform upon Eq. (8.1) as follows:

$$h(r) = 2 \int_0^{2\pi} \int_0^{\rho_0} P(\rho, \theta) \exp\left\{-\frac{j2\pi}{\lambda f}(\rho r \cos\theta)\right\} \rho \, d\rho \, d\theta. \tag{8.2}$$

r is the radial coordinate in the Fourier plane (x, y).

The Straubel aperture is independent of the angle θ. It is varied with the radial coordinate ρ.

By substituting Eq. (8.1) with Eq. (8.2), we write the following:

$$h(r) = 4\pi \int_0^{\rho_0} \left[1 - \beta\left(\frac{\rho}{\rho_0}\right)^2\right] \cdot J_0\left(\frac{2\pi}{\lambda f}\rho r\right) \rho \, d\rho, \tag{8.3}$$

where J_0 is the Bessel function of zero order.

We rewrite Eq. (8.3) as follows:

$$h(r) = 4\pi (I_1 - I_2), \tag{8.4}$$

$$I_1 = \int_0^{\rho_0} \rho J_0\left(\frac{2\pi}{\lambda f}\rho r\right) d\rho, \tag{8.5}$$

$$I_2 = \beta \int_0^{\rho_0} \rho\left(\frac{\rho}{\rho_0}\right)^2 J_0\left(\frac{2\pi}{\lambda f}\rho r\right) d\rho. \tag{8.6}$$

We replace the reduced coordinate $w = \frac{2\pi}{\lambda f}\rho r$ instead of the radial coordinate r in Eqs. (8.5 and 8.6), then we write the following:

$$I_1 = \left(\frac{\lambda f}{2\pi r}\right)^2 \int_0^W w J_0(w)\,dw; \quad W = \frac{2\pi r}{\lambda f}\rho_0, \tag{8.7}$$

$$I_2 = \left(\frac{\beta}{\rho_0^2}\right)\left(\frac{\lambda f}{2\pi r}\right)^4 \int_0^W w^3 J_0(w)\,dw, \tag{8.8}$$

$$I_1 = \left(\frac{\lambda f}{2\pi r}\right)^2 [W \cdot J_1(W)] = (\rho_0^2)\cdot\frac{J_1(W)}{W}, \tag{8.9}$$

$$I_2 = \left(\frac{\beta}{\rho_0^2}\right)\left(\frac{\lambda f}{2\pi r}\right)^4 \left[W^3 J_1(W) - 2W^2 J_2(W)\right]; \quad \frac{\lambda f}{2\pi r} = \frac{\rho_0}{W}$$

$$I_2 = \beta\rho_0^2 \left[\frac{J_1(W)}{W} - 2\frac{J_2(W)}{W^2}\right]. \tag{8.10}$$

By substituting Eqs. (8.9 and 8.10) in Eq. (8.4), we obtained the PSF corresponding to the Straubel aperture as follows:

$$h(W) = 4\pi\rho_0^2 \left[\frac{J_1(W)}{W}\right] - 4\pi\beta\rho_0^2 \left[\frac{J_1(W)}{W} - 2\frac{J_2(W)}{W^2}\right]. \tag{8.11}$$

The PSF can be rewritten as follows:

$$h(W) = 4\pi\rho_0^2(1-\beta)\left[\frac{J_1(W)}{W}\right] + 8\pi\beta\rho_0^2\left[\frac{J_2(W)}{W^2}\right]. \tag{8.12}$$

For $\beta = 0.6$, Eq. (12) is reduced to:

$$h(W) = \left[\frac{J_1(W)}{W}\right] + 3\left[\frac{J_2(W)}{W^2}\right]. \tag{8.13}$$

In Eq. (8.13), we omitted the multiplicative term which represents the area of a circle $= 4\pi\rho_0^2$.

8.2.2 *Formation of Fourier Holograms Using Apertures as Objects*

The aperture is numerically recorded as a Fourier hologram using MATLAB code as follows: First, a diffuser of the same dimensions as the rescaled modulated aperture of 512×512 pixels is numerically constructed. Second, the product of both the modulated aperture and the diffuser is Fourier transformed numerically to obtain the holographic interference image. This hologram is the convolution product of the Fourier transforms of the aperture and the speckle pattern. The speckle distribution is the Fourier transform of the diffuser function. Third, the inverse Fourier transform is applied to the described Fourier hologram to obtain the reconstructed images of the aperture.

The two Fourier transform operations are summarized as follows:

First step (Recording the Amplitude and Phase of the Fourier Hologram)

The complex amplitude of the aperture is analytically represented as:

$$A(x, y) = a(x, y) \exp\!\big[j\Phi(x, y)\big]. \tag{8.14}$$

The image is numerically written as follows:

$$A(x, y) = \sum_{m=1}^{M} \sum_{n=1}^{N} a(m\Delta x, n\Delta y) \exp\!\big[j\Phi(m\Delta x, n\Delta y)\big], \tag{8.15}$$

where the continuous variables (x, y) are replaced by numerical values as follows:

$$x = m\,\Delta x, \quad \text{and} \quad y = n\Delta y; \quad j = \sqrt{-1}.$$

The digital diffuser is written as follows:

$$D(x, y) = \sum_{m=1}^{M} \sum_{n=1}^{N} \mathrm{rand}(m\Delta x, n\Delta y), \tag{8.16}$$

where rand is a random value between 0 and 1, and $\mathrm{rand}(m\Delta x, n\Delta y)$ is the amplitude weighting factor.

The complex amplitude of the Fourier hologram is obtained by operating the FFT upon the multiplication product of the two matrices represented in Eqs. (8.15) and (8.16).

Hence, the complex amplitude of the hologram $B(u, v)$ is represented as follows:

$$B(u, v; \lambda) = \text{F.T.}\big[A(x, y).D(x, y)\big]$$

$$= \text{F.T.}\{A(x, y)\} \otimes F.T.\{D(x, y)\} \xrightarrow{\text{yields}} A_{\text{holo}}(u, v) \otimes D_{\text{holo}}(u, v), \quad (8.17)$$

where (u, v) are the spatial coordinates in the Fourier plane.

Second Step (Reconstruction Process of the Inverse Fourier Transform)

$$c(x', y') = \text{F.T}^{-1}[A_{\text{holo}}(u, v) \otimes D_{\text{holo}}(u, v)]$$

$$= A_{\text{reconst.}}(x', y') \cdot D_{\text{reconst}}(x', y'), \quad (8.18)$$

where (x', y') represent the Cartesian coordinates in the imaging plane or the reconstruction plane. The numerical reconstruction images have maximum dimensions of $x' = x_{\max} = M \Delta x'$ and $y' = y_{\max} = N \Delta y'$ For the matrix of dimensions 1024 × 1024 pixels.

8.3 Results and Discussion

We computed the PSF curve corresponding to the Straubel aperture compared to the PSF curves in the cases of circular and quadratic apertures shown in Fig. 8.1a–c.

Referring to Fig. 8.1a, we showed that the PSF corresponding to the Straubel aperture has a full width at half maximum (FWHM) = 109 μm. Meanwhile, the FWHM = 101 μm is shown in Fig. 8.1b in the case of circular aperture. The FWHM = 93 μm is shown in Fig. 8.1c in the case of quadratic aperture.

It appears that the Straubel aperture may be considered like the inverted quadratic aperture. The PSF cut-off in the case of Straubel aperture is greater than that obtained in the case of circular and quadratic apertures. Referring to Tables 8.2, 8.3, and 8.4, the cut-off spatial frequency for the aperture radius = 32 pixels has the following inequality:

$w_c(\text{Quad.}) = 42\text{pixels} < w_c(\text{Circle}) = 44\text{pixels} < w_c(\text{Straubel}) = 45\text{pixels}$.

In Tables 8.2, 8.3, and 8.4, we showed that the contrast corresponding to the reconstructed images from the hologram is decreased with increasing aperture radius. Conversely, the resolution improves with the increase in aperture radius, as the cut-off spatial frequency obtained from the PSF decreases.

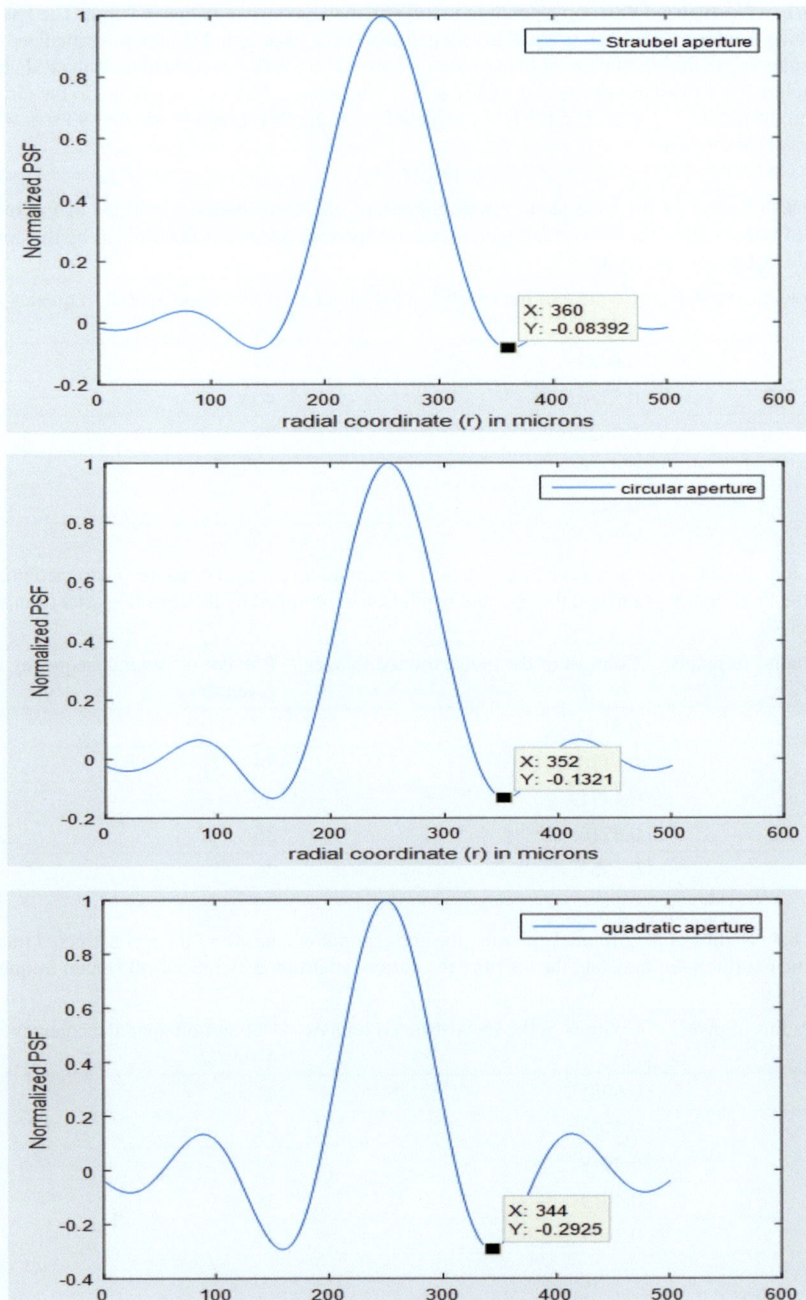

Straubel aperture
Normalized PSF
radial coordinate (r) in microns
X: 360
Y: -0.08392

circular aperture
Normalized PSF
radial coordinate (r) in microns
X: 352
Y: -0.1321

quadratic aperture
Normalized PSF
radial coordinate (r) in microns
X: 344
Y: -0.2925

◀**Fig. 8.1** **a** Normalized PSF corresponding to the Straubel aperture of radius = 2 mm. The FWHM computed from the PSF graph is equal to 109 μm where the peak is at 251 μm. **b** Normalized PSF corresponding to the circular aperture of radius = 2 mm. The FWHM computed from the PSF graph is equal to 101 μm where the peak is at 251 μm. **c** Normalized PSF corresponding to the circular aperture of radius = 2 mm. The FWHM computed from the PSF graph is equal to 93 μm where the peak is at 251 μm

Table 8.2 Radius of the transparent circular aperture, the corresponding contrast of the reconstructed images from the Fourier hologram, and the aperture resolution determined by the cut-off spatial frequency in pixels/λf

The radius in pixels	Contrast of the reconstructed images	PSF cut-off spatial frequency in pixels/λf
32	0.6483	44
48	0.5336	40
64	0.4622	39
96	0.3754	37
128	0.3228	36

Table 8.3 Radius of the quadratic aperture, the corresponding contrast of the reconstructed images from the Fourier hologram, and the aperture resolution determined by the cut-off spatial frequency in pixels/λf

The radius in pixels	Contrast of the reconstructed images	PSF cut-off spatial frequency in pixels/λf
32	0.6187	42
48	0.5118	39
64	0.4471	38
96	0.3716	36
128	0.3263	36

Table 8.4 Radius of the Straubel aperture, the corresponding contrast of the reconstructed images from the Fourier hologram, and the aperture resolution determined by the cut-off spatial frequency in pixels/λf

The radius in pixels	Contrast of the reconstructed images	PSF cut-off spatial frequency in pixels/λf
32	0.6981	45
48	0.5724	41
64	0.4958	39
96	0.4019	38
128	0.3427	37

We computed the Fourier hologram of the following objects. The first is a transparent circle, the second has a quadratic distribution, and the third has an inverted quadratic distribution, known as the Straubel filter. Each of the apertures has a radius = 128 pixels in the matrix of dimensions 512×512 pixels and uses a diffuser of the same dimensions as the objects. We showed the holograms and the corresponding reconstruction of images in a matrix of dimensions 1024×1024 pixels in Figs. 8.2, 8.3, and 8.4. Improved reconstruction of the Fourier holograms is obtained in Fig. 8.5a using the phase Fourier holography. Hundred iterations are shown in Fig. 8.5b.

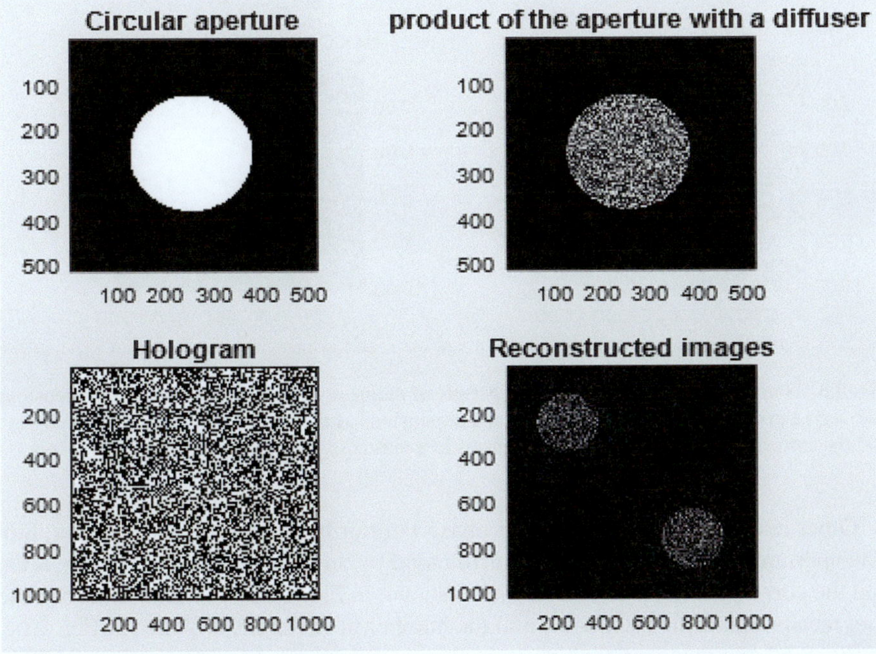

Fig. 8.2 Fourier hologram of a transparent circle of radius = 128 pixels in a matrix of dimensions 512×512 pixels using a diffuser of the same dimensions as the object. We showed the hologram and the corresponding reconstruction of images in a matrix of dimensions 1024×1024 pixels

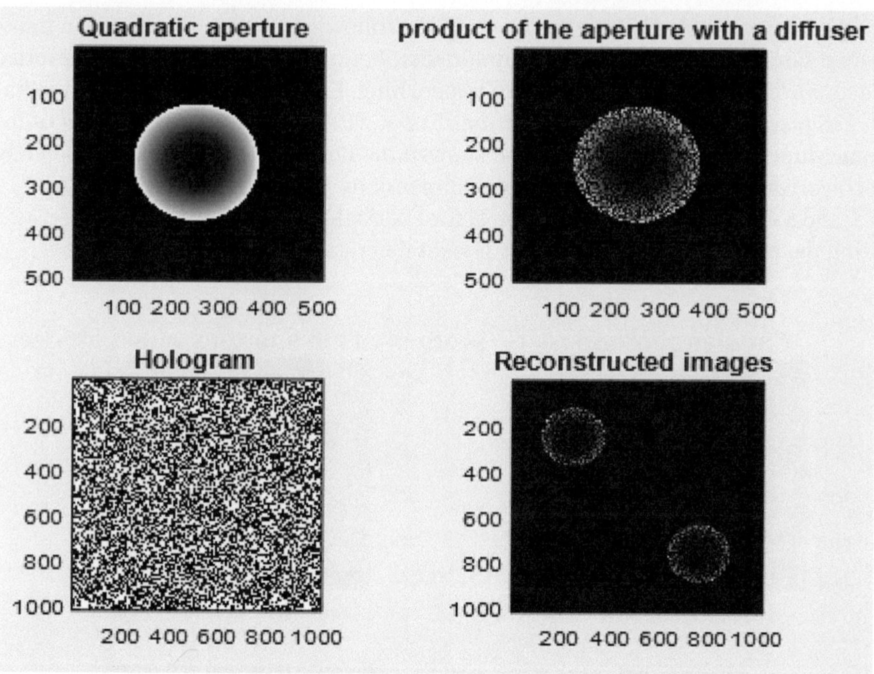

Fig. 8.3 Fourier hologram of a quadratic aperture of radius = 128 pixels in a matrix of dimensions 512 × 512 pixels using a diffuser of the same dimensions as the object. We showed the hologram and the corresponding reconstruction of images in a matrix of dimensions 1024 × 1024 pixels

Other results for the formation of phase Fourier hologram are given in Fig. 8.6. The aperture of eight linear circles surrounded by an annulus is shown in Fig. 8.6a, and the corresponding phase hologram is shown in Fig. 8.6b. A relation between the root means square error (RMSE) and the number of iterations is shown in Fig. 8.6c. In Fig. 8.6d, the reconstructed image is shown. All images have dimensions of 2048 × 2048 pixels.

8.3.1 Conclusion

We showed that the reconstructed images for different apertures as objects have a contrast increased with decreased aperture radius. In addition, the resolution computed from the cut-off spatial frequency obtained from the PSF is improved with the increase of the aperture radius. Consequently, the contrast is inversely proportional to the resolution governed by the aperture.

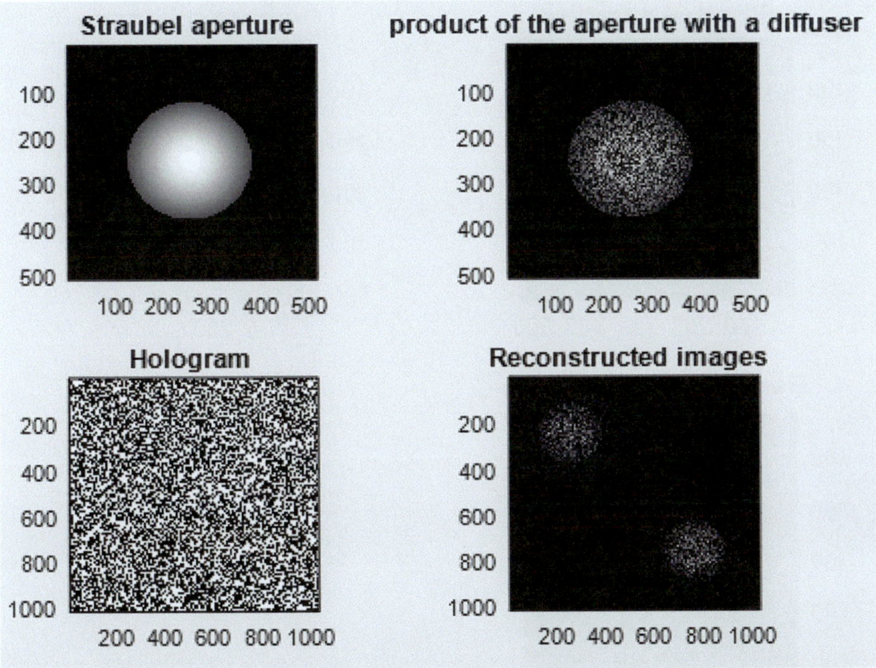

Fig. 8.4 Fourier hologram of Straubel aperture of radius = 128 pixels in a matrix of dimensions 512 × 512 pixels using a diffuser of the same dimensions as the object. We showed the hologram and the corresponding reconstruction of images in a matrix of dimensions 1024 × 1024 pixels

The Straubel aperture has FWHM greater than that corresponding to the circular and quadratic apertures. We may consider the Straubel aperture as the inverted quadratic aperture. We obtained moderate resolution in the case of the Straubel aperture compared to the circular and quadratic apertures. In addition, the Straubel aperture has a PSF that corresponds to the inverted quadratic aperture for the parameter $\beta = 1$.

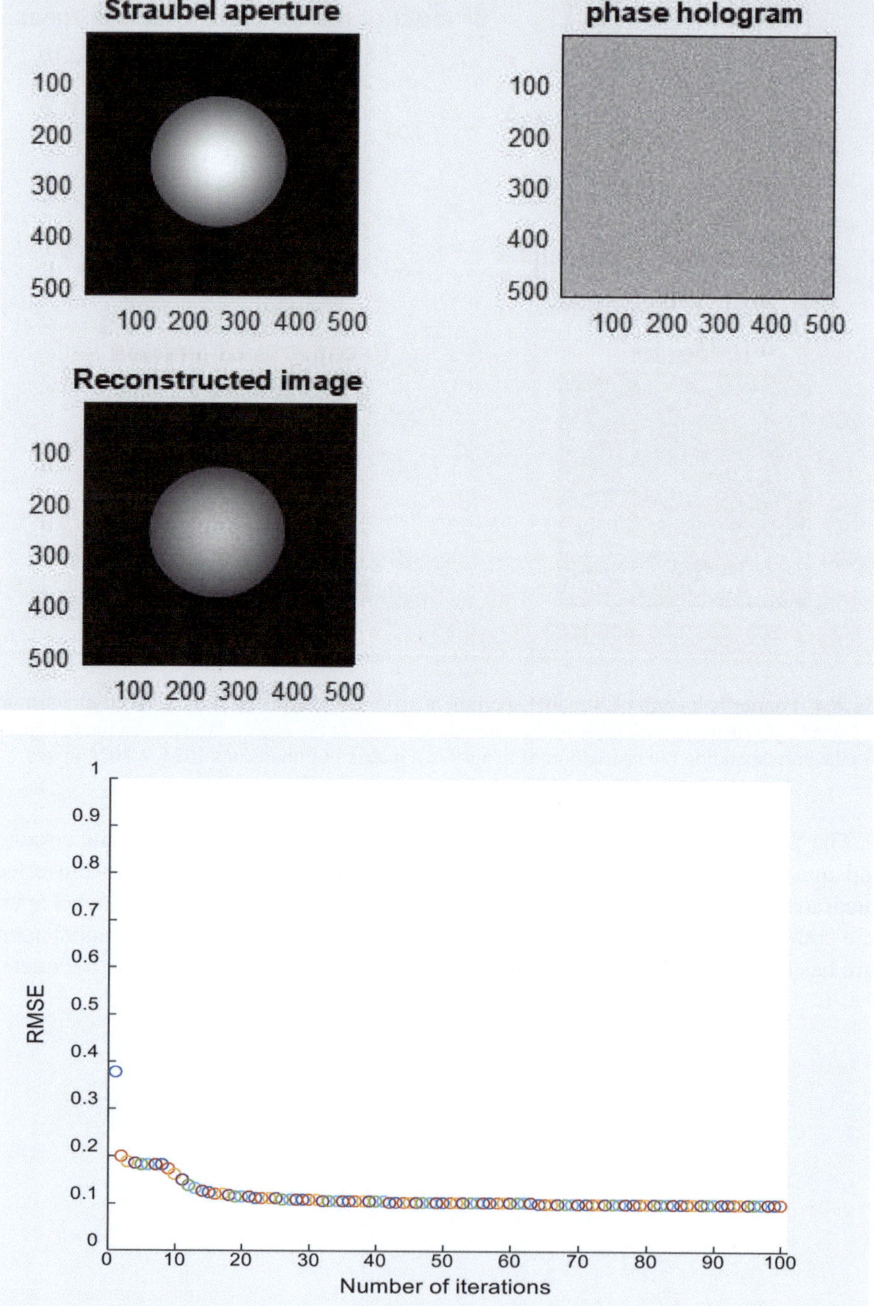

Fig. 8.5 **a** Construction of a phase Fourier hologram and improved reconstructed image corresponding to the Straubel aperture as an object. **b** Relation between the root means square error (RMSE) versus the number of iterations that is shown in figure. 100 Iterations are made to obtain the reconstructed image shown in (**a**)

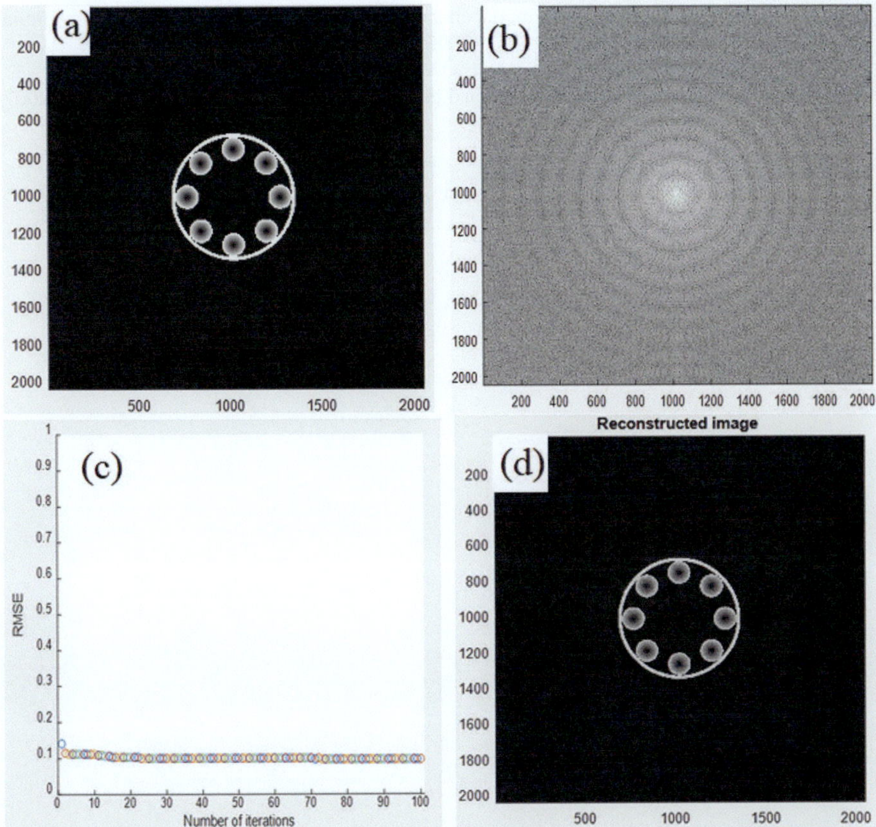

Fig. 8.6 Aperture of eight linear circles surrounded by an annulus is shown in (**a**) and the corresponding phase hologram is shown in (**b**). A relation between the root means square error (RMSE) and the number of iterations is shown in (**c**). In **d**, the reconstructed image is shown. All images have dimensions of 2048 × 2048 pixels

References

1. Stern, B. Javidi, Improved-resolution digital holography using the generalized sampling theorem for locally band-limited fields. J. Opt. Soc. Am. **A23**, 1227–1235 (2006)
2. L. Yu, Y. An, L. Cai, Numerical reconstruction of digital holograms with variable viewing angles. Opt. Express **10**, 1250–1257 (2002)
3. I. Yamaguchi, and T. Zhang, Phase-shifting digital holography. Opt. Lett. **22**, 1268–1270 (1997)
4. S. Grilli, P. Ferraro, et al., In-situ visualization, monitoring and analysis of electric field domain reversal process in ferroelectric crystals by digital holography. Opt. Express **12**, 1832–1842 (2004)
5. B. Javidi, E. Tajahuerce, Three-dimensional object recognition by use of digital holography. Opt. Lett. **25** (2000), 610–612
6. M.D.S. Hernandes-Montes, C. Perez-Lopez, et al., Detection of biological tissue in gels using pulsed digital holography. Opt. Express **12**, 853–858 (2004)

7. N. DeMolay, I. DeMolay, Dynamic modal characterization of musical instruments using digital holography. Opt. Express **13**, 4812–4817 (2005)
8. J.W. Godman, *Introduction to Fourier optics* (McGraw-Hill Education, New York, 1996)
9. D Apostol, A. Sima, et al., Fourier transform digital holography. Proc. SPIE **6785**, 678522 (2007)

Chapter 9
Recognition of Colored Objects Using Thick Holographic Multiplexed Filters

The thick holographic multiplexed filter described in this chapter simultaneously operates as a Fourier hologram and volume hologram. This filter was applied, in our studies, to the recognition of transparent colored objects by using the nominal double diffraction arrangement, which was illuminated with polychromatic light.

9.1 Introduction

Volume holograms are holograms where the thickness of the recording material is much larger than the light wavelength used for recording. In this case, the diffraction of light from the hologram is possible only as Bragg diffraction, i.e., the light must have the right wavelength (color) and the wave must have the right shape (beam direction, wavefront profile). Volume holograms are also called thick holograms or Bragg holograms. The volume hologram displays three-dimensional diffraction characteristics described earlier by the well-known Bragg relation $2d \sin(\theta) = n\lambda...$, where the spacing between reflecting planes is d, θ is the angle of incidence, and λ is the wavelength.

The use of thick media for making thick (volume) holograms was first proposed by DENISYUK [1], and then Stroke et al. [2] extended the technique to record reflection holograms in thick emulsions and reconstructed monochromatic images using white light.

We remember that the Vander–Lugt filter [3] is usually considered the base in the construction of any holographic filter. CASE [4] described a holographic multiplexed filter using polychromatic light. Wavelength-multiplexed matched spatial filters are used in an incoherent optical pattern recognition system. The system produces color-coded correlation responses that allow the simultaneous recognition of several objects and, in addition, reduces the requirements on filter positioning devices [4]. Recently,

© The Author(s), under exclusive license to Springer Nature Switzerland AG 2025

A. Hamed, *Holographic Imaging Using Aperture Modulation*,

SpringerBriefs in Applied Sciences and Technology,

https://doi.org/10.1007/978-3-031-96989-8_9

multiplexed color prints have been created by encoding information in two independent dimensions of elongated metal nanostructures, allowing for two different images to be read out under orthogonal polarizations of light [5, 6]. Metasurface holograms comprise subwavelength meta-atoms requiring advanced nanofabrication techniques to achieve high-quality fabrication performance. For instance, to fabricate multilayer metasurfaces, one could do multilayer optical/e-beam lithography [7–9] or align stacking with the transfer technology that has been widely used in 2D materials [10].

In this chapter, a thick holographic multiplexed filter (THMF) is described from experimental and theoretical points of view [11]. Optical correlation was employed for the recognition of colored shapes using polychromatic light for illumination.

9.2 Theoretical Analysis

The transmission amplitude function of the colored transparent shape shown in Fig. 9.1 can be represented by

$$g(x, y, \lambda) = \sum_{i=B,G}^{R} g_i(x, y, \lambda_i), \tag{9.1}$$

where λ_i is the polychromatic wavelength of light, $g_{i=B}$ is the spectral blue component of the colored shape, etc. The amplitude transmitted from the object, given by Formula (9.1), is Fourier transformed using a converging lens L to obtain the following equation in the holographic zone.

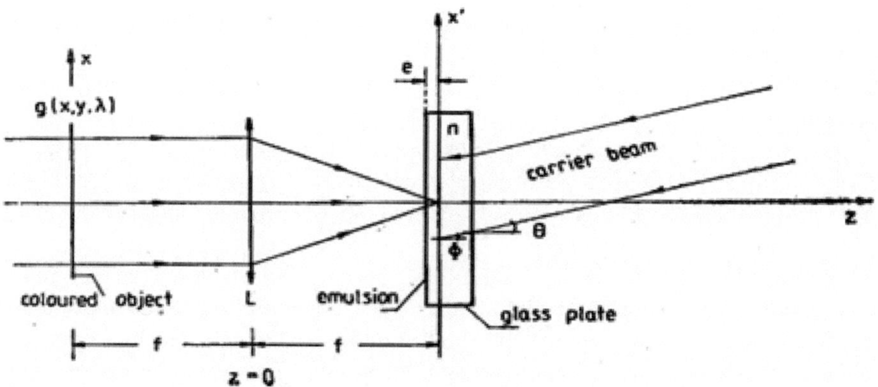

Fig. 9.1 Schematic diagram of the construction of a thick holographic multiplexed filter. $g(x, y, \lambda)$: colored object in the plane (x, y). L: Fourier transformed lens of focal length f. $H(x', y')$: holographic plane. n: refractive index of the glass plate. e: emulsion thickness

$$\text{F.T.}\{g(x, y, \lambda)\} = \text{F.T.}\left\{ \sum_{i=B,G}^{R} g_i(x, y, \lambda_i) \right\} = \sum_{i=B,G}^{R} \beta_i \tilde{g}_i(u, v, \lambda_i) \exp\left(jk_i \frac{r^2}{2z} \right).$$

$$(9.2)$$

For any color component, the Fourier spectrum $\tilde{g}_i(u, v, \lambda_i)$ is multiplied by an exponential quadratic term, indicating that we deal with a spatial variable plane inside the emulsion. $r = \sqrt{x'^2 + y'^2}$, $k_i = \frac{2\pi}{\lambda_i}$ is the wavenumber, and $u = \frac{x'}{\lambda f}$, $v = \frac{y'}{\lambda f}$ are the reduced coordinates in the plane (x', y'). The coefficient β_i depends mainly on the spectral intensity curve of the emulsion. $\sim$ denotes the performed transformation.

An inclined plane wave, with an angle θ with the z-axis incident on the opposite side of the holographic plate, is represented as follows:

$$A(x', y', \lambda) = \sum_i A_i \exp(-jk_i x' \sin \theta), \tag{9.3}$$

where A_i is the complex amplitudes that correspond to each spatial color component, and $k_i = \frac{2\pi}{\lambda_i}$ is the wavenumber corresponding to each wavelength of light illumination.

According to Snell's law of refraction and using Fig. 9.1, we obtain:

$$\sin \theta = n \sin \phi,$$

where the refractive index of the glass plate is n. Hence, the incident carrier plane wave that makes an angle ϕ is written as follows:

$$A(x', y', \lambda) = \sum_i A_i \exp\left[(-jk_i x' \sin \theta)/n \right]. \tag{9.4}$$

From Equations (9.2) and (9.4), we write the intensity recorded inside the holographic emulsion as follows:

$$I(x', y', z, \lambda) = \left| \sum_{i=B,G}^{R} \beta_i \tilde{g}_i(u, v, \lambda_i) \exp\left(jk_i \frac{r^2}{2z} \right) + \sum_i A_i \exp\left[(-jk_i x' \sin \theta)/n \right] \right|^2.$$

$$(9.5)$$

Assuming that the polychromatic light components are mutually incoherent and that the colored information is separated spatially, we can write Eq. (9.5) as follows:

$$I(x', y', z, \lambda) = \left| \sum_{i=B,G}^{R} \beta_i \tilde{g}_i(u, v, \lambda_i) \right|^2 + \left| \sum_i A_i \right|^2$$

$$+ 2 \sum_i \beta_i A_i \tilde{g}_i(u, v, \lambda_i) \cos\left[k_i \left(x' \sin \phi + \frac{r^2}{2z} \right) \right]. \tag{9.6}$$

It should be noted that the hologram must be recorded in the focal plane of the converging lens to ensure that we have a Fourier hologram. This condition is implied as shown in Fig. 9.2.

$$f \gg \pi\left(\frac{a^2}{\lambda}\right), \quad \text{and} \quad r = a = \sqrt{x'^2 + y'^2},$$

where a is the transverse height of the Gaussian spot in the plane (x', y'). From the characteristics of the Gaussian laser beam, it follows that the radius of the laser convergent beam is

$$a = \left(\frac{\lambda f}{\pi d}\right),$$

where the contraction angle $\gamma = \frac{\lambda}{\pi a}$ and $d = \gamma f$. Hence, to record the Fourier hologram inside the emulsion of thickness e, the following condition must be satisfied:

$$e < \left(\frac{\lambda}{\pi}\right)\left(\frac{f}{d}\right)^2.$$

For example, if $\lambda = 0.63$ μm, $f = 50$ cm, and $d = 1$ cm, then $e < 500$ μm, which is practically admitted.

If the amplitude transmitted from the holographic plate after the development process is linearly proportional to the intensity recorded in the emulsion, i.e., $t \sim I$, from Eq. (9.6), we obtain:

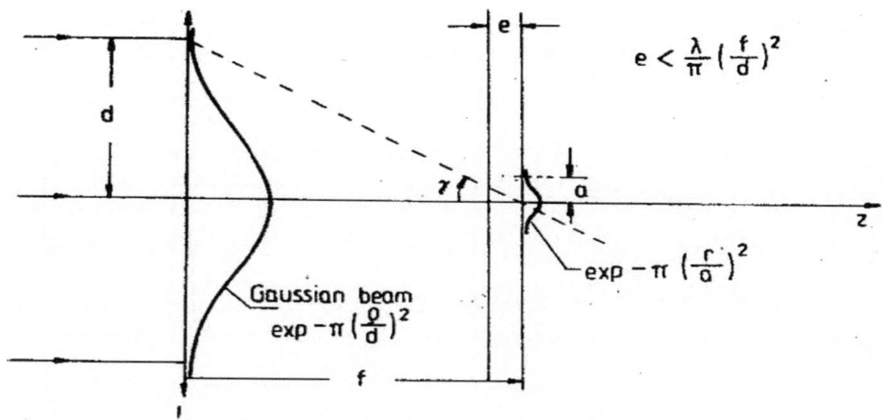

Fig. 9.2 Schematic diagram illustrating the necessary conditions to construct a Fourier hologram within an emulsion of thickness e. d: transversal height of the Gaussian beam in the pupil plane, a: transversal height of the Gaussian beam in the plane (x', y'), f: focal length of the lens L

$$t(x', y', z, \lambda) = \left| \sum_{i=B,G}^{R} \beta_i \tilde{g}_i(u, v, \lambda_i) \right|^2 + \left| \sum_i A_l \right|^2$$

$$+ 2 \sum_{i=l} \beta_i A_i \tilde{g}_i(u, v, \lambda_i) \cos\left[k_i \left(x' \sin \phi + \frac{r^2}{2z} \right) \right]. \qquad (9.7)$$

Consequently, the reconstructed image can be calculated by taking the Fourier transform of Eq. (9.7):

$$I_r(x, y, \lambda) = \text{F.T.}\{t(x', y', z, \lambda)\}, \quad r \to 0. \qquad (9.8)$$

Now, we analyze the correlator shown in Fig. 9.6. The amplitude transmitted from the plate multiplied by the Fourier transform of the amplitude transmitted from the object to the imaging plane should be Fourier transformed again.

$$A(x'', y'', \lambda) = \text{F.T.}\{t(x', y', z, \lambda) \times \text{F.T.}\{g(x, y, \lambda)\}\}. \qquad (9.9)$$

By performing the transformation in Eq. (9.9), we obtain:

$$A(x'', y'', \lambda) = \text{F.T.}\{t\} \otimes g(x'', y'', \lambda), \qquad (9.10)$$

where $\otimes$ denotes the convolution operation.

Since we are interested in the modulation term of interference in the analysis, we obtain the following from Eq. (9.10):

$$A(x'', y'', \lambda) = \text{F.T.}\left\{ \sum_{i=l} \beta_i A_i \tilde{g}_i(u, v, \lambda_i) \cos\left[k_i \left(x' \sin \phi + \frac{r^2}{2z} \right) \right] \right\} \otimes g(x'', y'', \lambda)$$

$$(9.11)$$

if $r^2 \ll \left(\frac{\lambda}{\pi}\right)z$, and $A_i = 1$. Then, after calculating the Fourier transformation, Eq. (9.11) gives the correlation amplitudes as follows:

$$A_c(x'', y'', \lambda) = g(x'', y'', \lambda) \otimes g(x'' + f \sin \phi, y'', \lambda). \qquad (9.12)$$

The correlation intensity detected in the imaging plane is calculated from Eq. (8.12) as follows:

$$I_c(x'', y'', \lambda) = \mid A_c(x'', y'', \lambda) \mid^2. \qquad (9.13)$$

For two colored (red and green) spatially separated objects, the correlation intensity is expressed as follows:

$$I_c = \mid g_R \otimes g_R \mid^2 + \mid g_G \otimes g_G \mid^2.$$

There is no cross-correlation term between the green and red components since they are spatially separated and because of the function of the thick holographic multiplexed filter.

9.3   Experiment

First, we describe the experimental arrangement used for the construction of the thick holographic multiplexed filter shown in Fig. 9.3, where the holographic grating results from the interference of the Fourier spectral components of the colored signal, obtained using a converging lens L and inclined carrier plane wave incident on the opposite side of the holographic plate. We simultaneously reconstruct both the shape of the object and its color utilizing the setup in transmission Fig. 9.4 or at the reflection as in Fig. 9.5.

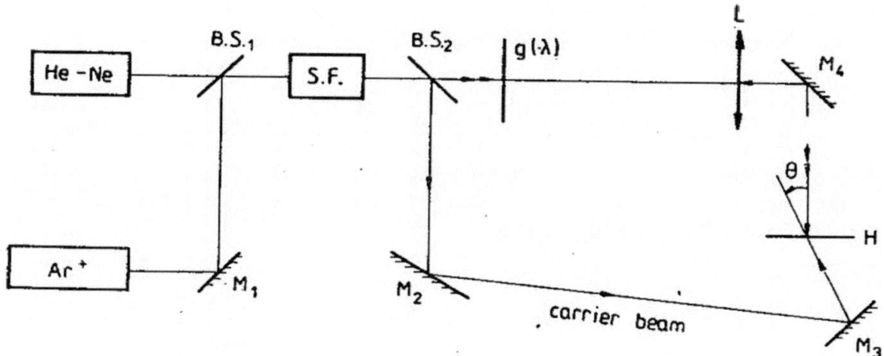

Fig. 9.3 Experimental arrangement for THMF construction: *S. F.*: spatial filter, $g(\lambda)$: colored object, *B. S.*: beam splitter, *M*: mirror, *H*: holographic domain, f: focal length of the lens *L*, which is the distance between *L* and H through M4

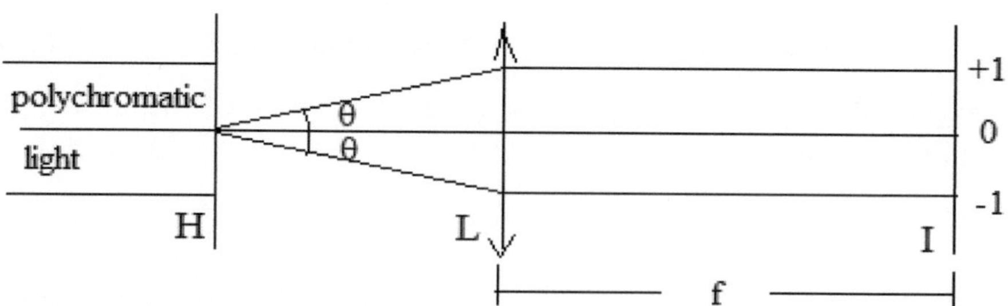

Fig. 9.4 Reconstruction of the colored image at transmission: *H*: processed hologram, *L*: Fourier transform lens of focal length f, *I*: image plane

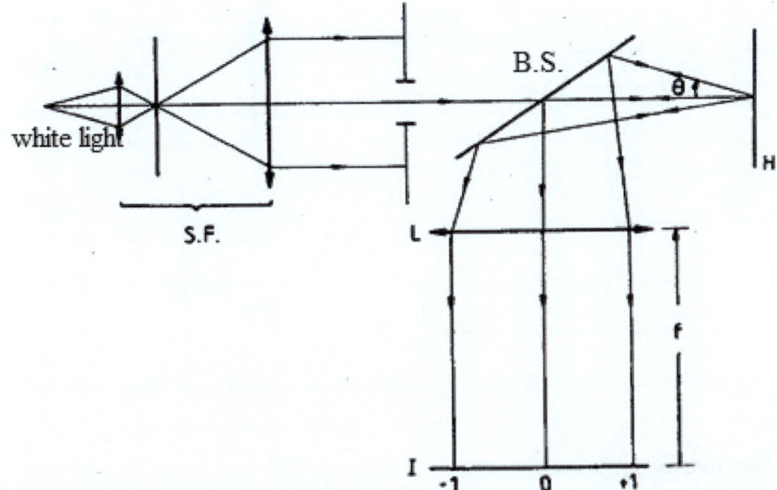

Fig. 9.5 Reconstruction system at reflection: *S. F.*: spatial filter, *B. S.*: beam splitter, *H, L, I* defined in Fig. 8.4

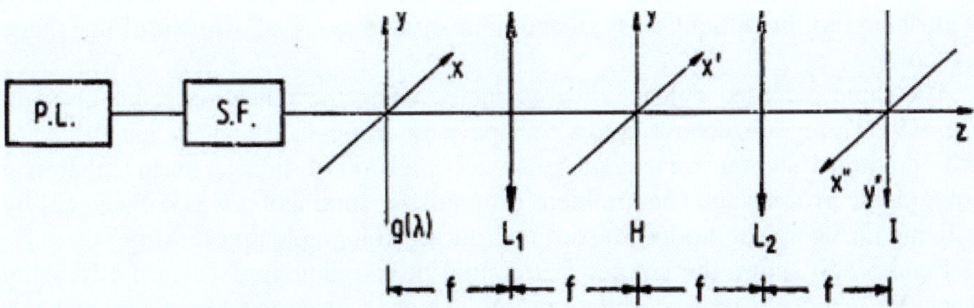

Fig. 9.6 Optical polychromatic correlator at transmission: *P. L.* polychromatic light, $L_{1,2}$: Fourier transform lenses with a focal length of 40 cm, $g(\lambda)$ Colored object, $H(x', y')$: processed photographic plates, and $I(x'', y'')$: reconstruction plane

Second, we describe the schematic arrangement of double diffraction using polychromatic light emitted from the He–Ne laser and Ar ion laser, which works in transmission, as shown in Fig. 9.6, and in reflection, as shown in Fig. 9.7. In both optical systems, the developed holographic plate is found in the back focal plane of the lens *L1*. The optical system working at reflection has an amplitude reflected from the developed holographic plate given by $r = \sqrt{1 - |t|^2}$, assuming that the absorption loss is negligible, and t is the transmitted amplitude. In the (x'', y'') imaging plane, colored correlation peaks are detected (Figs. 9.6 and 9.7).

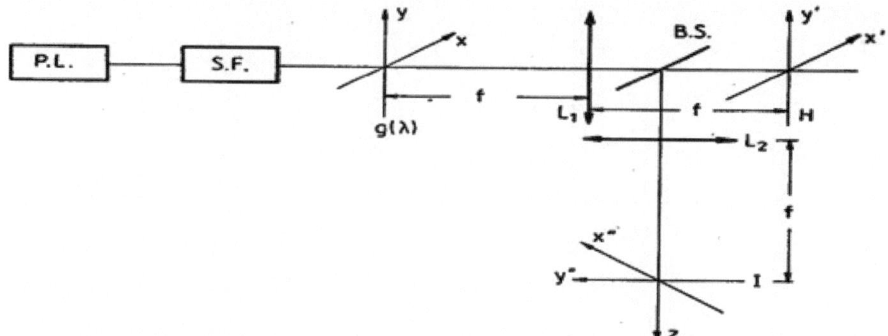

Fig. 9.7 Optical polychromatic correlator at reflection (we used the same notation as in Fig. 9.6)

9.4   Results and Discussion

Two colored letters (I is red and *V* is green) are recorded on a thick emulsion, and the reconstructed image corresponding to these colored shapes is shown in Fig. 9.8. The images reconstructed via transmission on both sides are parasitic images, which is attributed to the difficulty associated with simultaneous adjustment of the shape and color.

The same recorded object is reconstructed using the reflection setup given in Fig. 9.9. We reconstructed the exact shape without parasite images, but the color shifted toward shorter wavelengths due to emulsion shrinkage during chemical holographic processing. This problem of emulsion contraction was solved [12] by utilizing the chemical product D sorbitol* during holographic processing.

Figure 9.10 shows the colored correlation peaks photographed in the imaging plane. We detected a green correlation peak, which is represented by the green letter *V*, and we detected a red correlation peak corresponding to the letter *I* (red). If the color or shape of the object is modified, the correlation/peaks vanish, and the

Fig. 9.8 Black and white photograph of the reconstruction at transmission for the colored image. The actual image is in the center, while parasite images are on both sides

Fig. 9.9 Black and white photograph of the reconstruction at reflection for the same-colored image

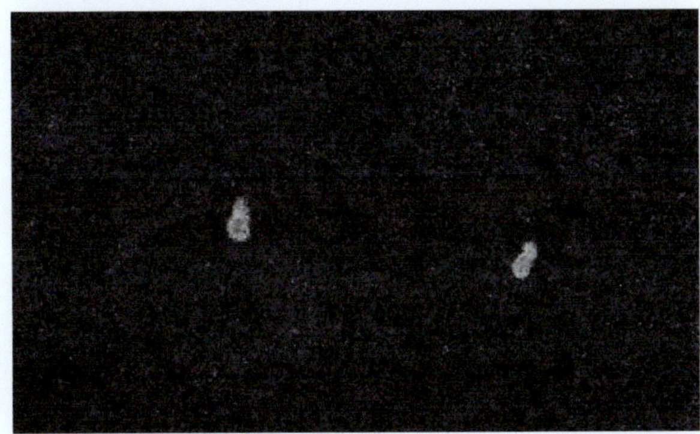

Fig. 9.10 Black and white photographs of the colored crrelation peaks. The green peak for the letter *V* is shown in the R.H.S., while the red peak for the letter *I* is shown in the L.H.S

detection fails. This confirms that real recognition is possible by using polychromatic correlation provided by a thick holographic multiplexed filter.

References

1. X. Yug, DENISYUK. Doklay Akkad. Nauk SSSR **144**, 1275 (1962)
2. G.W. Stroke, A.E. Labeyrie, Phys. Lett. **20**, 368 (1966)
3. J.W. Goodman, *Introduction to Fourier Optics* (McGraw Hill Book Co., New York, 1968)
4. S.K Case, Pattern recognition with wavelength-multiplexed filters. Appl. opt. **18**, 1890–1894 (1979)
5. X.M. Goh et al., Three-dimensional plasmonic stereoscopic prints in full color. Nat. Commun. **5**, 5361 (2015)
6. E. Heydari, J.R. Sperling, S.L. Neale, A.W. Clark, Plasmonic color filters as dual-state nano pixels for high-density micro image encoding. Adv. Funct. Mater. **27**, 1701866 (2017)
7. O. Avaya, E. Almeida, Y. Prior, T. Ellenbogen, Composite functional metasurfaces for multispectral achromatic optics. Nat. Commun. **8**, 1–7 (2017)
8. A. Arbabi, E. Arbabi, S. M. Kamali, Y. Horie, S. Han, A. Faraon, Miniature optical planar camera based on a wide-angle metasurface doublet corrected for monochromatic aberrations. Nat. Commun.**7**, 1−9 (2016)
9. B. Grover, W.T. Chen, F. Capasso, Meta-lens double in the visible region. Nano Lett. **17**, 4902–4907 (2017)

10. Y. Zhou, I.I. Kravchenko, H. Wang, H. Zheng, G. Gu, J. Valentine, Multifunctional meta optics based on bilayer metasurfaces. Light Sci. Appl. **8**, 1–9 (2019)
11. A.M. Hamed, Recognition of colored objects using thick holographic multiplexed filter. Opt. Applicate **13**, 205–2013 (1983)
12. P Hariharan, Improved techniques for multicolor reflection holograms. J. opt. **11**, 53 (1980)

Chapter 10
Pattern Recognition and Processing of Information

In this chapter, we present the mathematical basics of optical filtering. In addition, we recognize colored plane objects of simple shape using holographic filters.

10.1 Introduction

The domain of coherent optics is due to Fresnel (beginning of the nineteenth century). The appearance of gas lasers in 1961 made important progress in the field of coherent optics. This field is based on the scalar theory of diffraction. Two important variants of laser are spatial coherence and temporal coherence relative to the classical sources. A considerable contribution due to Marechal is around spatial coherence filtering and image formation [1]. The holography imagined by Gabor [2, 3] then is improved by the group of Michigan University; Stroke and Leith et al. [4–8], and others [9–11]. However, it is possible to record the amplitude and phase information of an object by adding a carrier wave to the object field. The recording is made on a high-resolution photographic plate sensitive to the wavelength of light. The important properties of holograms are the following:

1. The reconstruction of a relief image is obtained for the recording of a three-dimensional object.
2. It is possible to record at different times many objects on the same emulsion.

Optical information processing has been a major field of research. The holographic techniques were found extremely useful in the synthesis of complex filters used in image restoration as well as matched spatial filters used in optical pattern recognition and character reading [12]. The development of holography was greatly stimulated by the application of the concepts of communication theory to optics. The fact that a multicolor image can be produced by a hologram recorded with three suitable chosen wavelengths was first pointed out by Leith and Upatnieks [6]. The resulting

135

A. Hamed, *Holographic Imaging Using Aperture Modulation*,
SpringerBriefs in Applied Sciences and Technology,
https://doi.org/10.1007/978-3-031-96989-8_10

recording can be considered as made up of three incoherently superposed holograms. Recognition of colored objects using a thick holographic multiplexed filter (THMF) is presented in [13].

In this chapter, Fourier holograms described in Chap. 2 are used in the processing of color information using a polychromatic correlator. A characteristic figure of recognition is deduced.

10.2　Mathematical Basics of Optical Filtering

10.2.1　The Correspondence Relation

Optical filtering may be justified through mathematical considerations using Fourier optics and convolution operations. It is required to find the formula that relates a two-dimensional signal $g(M_0)$. to the output response $r(M)$ obtained by a coherent optical system of an impulse response $h(M)$. It is sufficient to assume that the transformation is linear and that the impulse response of the instrument is invariant by translation, i.e., for a signal represented by a Dirac distribution $\delta(M_0)$, and if $\delta(M_0)$ as appoint object gives the response $h(M)$, then the impulse $\delta(M_0 - M_1)$ displaced by M_1 through translation gives the response $h(M - M_1)$.

The correspondence relation between the input signal and the output signal is written as follows; Goodman (1968):

$$r(M) = g(M) \otimes h(M), \tag{10.1}$$

where the symbol $\otimes$ represents the convolution operation. This symbolic representation of Eq. (10.1) is written in integral form as follows:

$$r(x, y) = \int_{-\infty}^{\infty} g(x - x', y - y')h(x', y')\mathrm{d}x'\mathrm{d}y', \tag{10.2}$$

where $g(M) = g(x, y)$, and $h(M) = h(x, y)$.

Equation (101.1) may be written in the frequency plane $\Omega\,(u, v)$ as follows:

$$R(\Omega) = \text{F.T.}[r(M)] = \text{F.T.}\big[g(M) \otimes h(M)\big]. \tag{10.3}$$

F.T. is the Fourier transformation operation. Since the Fourier transform of convolution is equal to the simple multiplication of the Fourier transform corresponding to each term, hence we obtain:

$$R(\Omega) = G(\Omega) \cdot H(\Omega), \tag{10.4}$$

where $G(\Omega) = \text{F.T.}\big[g(M)\big]$ and $H(\Omega) = \text{F.T.}\{\,h(M)\}$.

We deduced that the relation between $g(M_0)$ and $r(M)$ is linear. The schematic representation of spatial filtering is shown in Fig. 11.1. $\Omega\,(u, v)$ is a current point in the Fourier space. The above reasoning utilizes mathematical tools. It is applied equally to incoherent systems like coherent systems. The schematic treatment may be realized optically using a double diffraction arrangement. The signal diffracted by the lens L_0 gives the Fourier spectrum. This spectrum is filtered by employing the second diffraction giving the output response. In general, the signal will give in the spectral plane the product of the Fourier spectrum and the spherical phase term. To get linear filtering, we must compensate for this quadratic-phase factor.

We showed that linear filtering is obtained if the objective lens L of the reconstruction system has its pupil in the spectral plane. This is realized in three steps:

10.2.2 Spectral Analysis

When the coherent convergent wave is propagating through the objective lens L_0, the signal $g(M_0)$ in the plane Π_0 is diffracted in the plane P of a current point $p(x_1, y_1)$ giving the amplitude

$$W(\rho) = \exp\!\left(ik\,\frac{\rho^2}{2d}\right) \cdot G(\rho). \tag{10.5}$$

d is the distance between the plane Π_0 and P, $G(\rho) = \text{F.T.}\{g(M_0)\}$, and $\Omega(u, v) = \frac{\rho}{\lambda f}$; $\rho = (x_1, y_1)$.

10.2.3 Spatial Filtering

In the plane P, a filter of an amplitude $H(\Omega)$ is placed. In the special case, this filter may be simple. After traversing the filter, the complex amplitude of the wave $G(\Omega)$ becomes:

$$R(\Omega) = G(\Omega) \cdot H(\Omega). \tag{10.6}$$

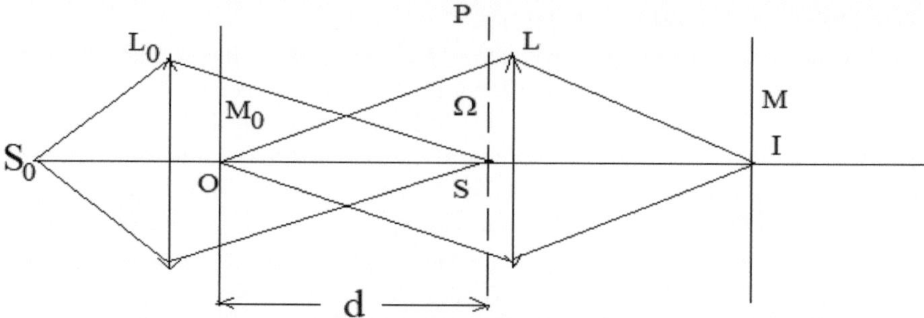

Fig. 10.1 Reconstruction setup using double diffraction arrangement to assure the linearity conditions. The pupil of the reconstruction objective L must be placed in the spectral plane P

10.2.4   Reconstruction Process

For the assurance of the above conditions of linearity and compensation of the phase factor, the objective lens (L) has its entrance pupil in the spectral plane P. This is schematically shown in Fig. 10.1.

Hence, for the assurance of linearity and translation invariance, the pupil of the objective L must be placed in the spectral plane P. The transverse magnification of L is normalized to $W = 1$. After traversing through L, the wave $R(\Omega)$ will give in the imaging plane I of a current point $M(x, y)$ in the following response:

$$R(M) = \int G(\Omega) \cdot H(\Omega) \exp\left[2\pi i\left(\overline{\Omega} \cdot \overline{M}\right)\right]\mathrm{d}\Omega, \tag{10.7}$$

which is the inverse Fourier transform of the filtered spectrum, i.e.,

$G(\Omega) \cdot H(\Omega) \rightarrow$ F.T.$^{-1}r(M)$, and the impulse response of the filter has a reciprocal relation with the filter function $H(\Omega) \rightarrow$ F.T.$^{-1}h(M)$. This response $h(M)$ is the image of a point object placed at the origin of the plane Π_0. Applying the convolution theorem to Eq. (10.4), we get the response function as follows:

$$r(M) = \text{F.T.}\{G(\Omega) \cdot H(\Omega)\} = g(M) \otimes h(M). \tag{10.8}$$

This is summarized as follows:

$$g(M) \xrightarrow{\text{F.T.}} G(\Omega) \longrightarrow \otimes \longrightarrow R(\Omega) = G(\Omega)\,H(\Omega) \longrightarrow r(M) = g(M)*h(M)$$

$$\text{signal} \qquad\qquad \text{spectrum} \qquad\qquad\qquad \text{filtered spectrum} \qquad\qquad \text{response}$$

$$H(\Omega) \xrightarrow{\text{F.T.}} h(M)$$

$$\text{filter}$$

10.3 Recognition of an Object Using a Holographic Filter

Assume a colored plane object of a simple shape and colors. We discuss a holographic filter adapted to the recognized object for a certain wavelength. We say that a linear invariant spatial filter is adapted to the signal $S(x, y)$ if for a certain wavelength λ the impulse response $h_\lambda(x, y)$ is written as follows: $h_\lambda(x, y) = S_\lambda(-x, -y)$.

The recorded hologram is considered as the Fourier transform of the impulse response, i.e.,

$$H_\lambda(f_x, f_y) = \text{F.T.}_\lambda\big[h(x, y)\big]$$

The formation of a holographic adapted filter is discussed in Chap. 2 in Sect. 2.7.

10.4 Two-Dimensional Polychromatic Correlator

First, we start to describe briefly the monochromatic correlator using a double diffraction arrangement as shown in Fig. 10.2. The complex amplitude of the object $g(x_1, y_1)$ placed in the input plane is Fourier transformed by the lens L_1 obtaining in the following spectrum:

$$\tilde{g}(\xi, \eta) = \text{F.T.}\big[g(x_1, y_1)\big].$$

The processed photographic plate of the Fourier hologram recorded in the previous section is multiplied by the above spectrum, to yield this result in the Fourier plane (ξ, η) [14]:

$$
\begin{aligned}
B(\xi, \eta) = {}& \tilde{g}(\xi, \eta) \\
& \times \big[\tilde{g}^2 + R^2 + R^* \cdot \tilde{g}\exp(j2\pi\alpha x_2) + R \cdot \tilde{g}^*\exp(-j2\pi\alpha x_2)\big],
\end{aligned}
\qquad (10.9)
$$

where $\alpha = \frac{\sin(\theta)}{\lambda}$.

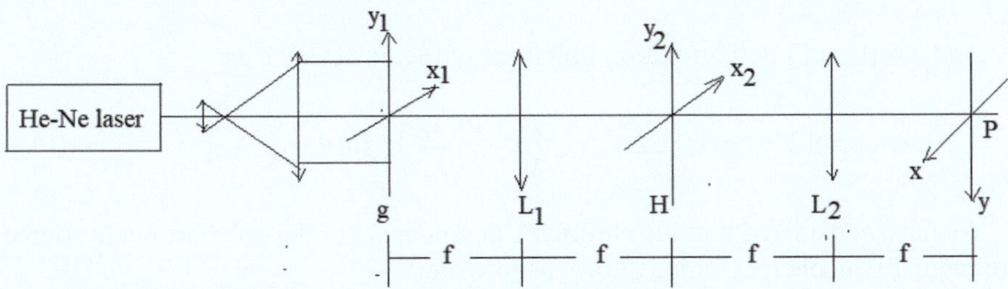

Fig. 10.2 Double diffraction arrangement using monochromatic light. $g(x_1, y_1)$: object plane, $H(x_2, y_2)$: spectral plane of the hologram, and $P(x, y)$: imaging plane

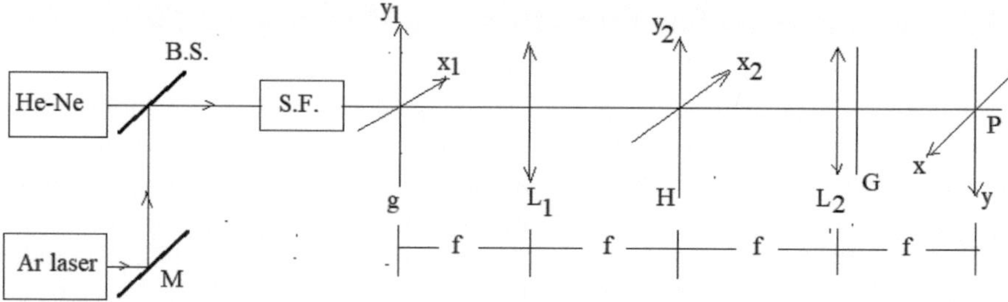

Fig. 10.3 Double diffraction arrangement using polychromatic light. S.F.: spatial filter, B.S.: beam splitter, M: mirror, $g(x_1, y_1)$: object plane, L_1, L_2: converging lenses, $H(x_2, y_2)$: spectral plane of the hologram, and $P(x, y)$: imaging plane

Neglecting the first two terms that appeared inside the square bracket, we obtain:

$$B(\xi, \eta) = \tilde{g}(\xi, \eta) \cdot R^* \cdot \tilde{g} \exp(j2\pi\alpha x_2) + \tilde{g}(\xi, \eta) \cdot R.\tilde{g}^* \exp(-j2\pi\alpha x_2). \quad (10.10)$$

Consequently, operating the inverse Fourier transform by using the lens L_2, we obtain the following in the output plane (x, y):

$$C(x, y) = g(x, y) \otimes g(x - f \sin\theta, y) + g(x, y) \otimes g^*(x + f \sin\theta, y), \quad (10.11)$$

where R is set equal to unity.

Now, we study the image obtained in the output plane in the case of polychromatic illumination. Referring to Fig. 10.3, we place the holographic filter adapted to the recognized object at a certain wavelength λ.

In the following, we assume a colored object composed of monochromatic object components λ_1 found between the wavelength λ_i. It should be noted that the hologram plays an important role as a dispersive element and the presence of a supplementary dispersive element like grating increases the dispersive power of the instrument. In all cases, the measured intensity in the plane P gives the autocorrelation of the object as:

$$I(\text{autocorrel.}) = \mid g(x, y, \lambda_1) \otimes g^*(-x + f \sin\theta, -y, \lambda_1) \mid^2 . \quad (10.12)$$

And a series of intercorrelation functions of the object such as:

$$I(\text{intercorrel.}) = \mid g(x, y, \lambda_i) \otimes g^*\left(x + \left(\frac{\lambda_i}{\lambda_1}\right)f \sin\theta, -y, \lambda_i\right) \mid^2 . \quad (10.13)$$

We have considered that the chromatic components of the polychromatic source are mutually incoherent which allows us to write:

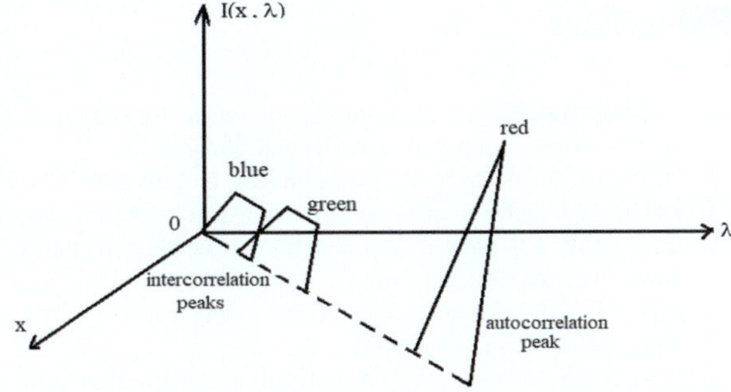

Fig. 10.4 Characteristic figure of recognition (CFR)

$$I(\text{intercorrel.}) = \sum_i \{I(\text{intercorrel.})\} = I(\lambda_1) + I(\lambda_2) + \cdots . \tag{10.14}$$

These results of correlation are represented in the following section.

10.5 Effect of the Dispersive Element upon Correlation

In the imaging plane, the intercorrelation intensity is represented by the summation represented in Eq. (10.14). If the summation is restricted to one wavelength, the above expression will reduce to the autocorrelation intensity, while if the summation is extended to different wavelengths λ_i, then the characteristic figure of intercorrelation of the object is obtained. Finally, if the summation is extended to all wavelengths λ_i including the wavelength λ_1, the obtained figure is called the characteristic figure of recognition (CFR), Hamed et al. [15] (Fig. 10.4).

The introduction of the dispersive element (e.g., diffraction grating) between the imaging lens and the imaging plane increases the separation distances between the different figures of correlation.

Utilizing the natural dispersity of the hologram, the distance between the first order and zero order (central image) in the imaging plane along the x-axis is: $x = f \sin(\theta)$, and for a certain wavelength λ_i, $x_i = \left(\frac{\lambda_i}{\lambda_1}\right).f \sin(\theta)$. If $\lambda_i = \lambda_1$, then $x = f \sin(\theta)$.

We deduce that the distance between the successive figures of intercorrelation is equal to:

$$\left|x_j - x_i\right| = \left|\frac{\lambda_j - \lambda_i}{\lambda_1}\right| f \sin(\theta)$$

References

1. A. Maréchal, Theory and practice of image formation. J. Opt. Soc. Am. **56**(12), 1645–1648 (1966). https://doi.org/10.1364/JOSA.56.001645
2. D. Gabor, A microscopic new principle. Nature **161**, 777–778 (1948)
3. D. Gabor, Dennis, microscopy by reconstructed wavefronts. Proc. R. Soc. **197**, 454–487 (1949)
4. E.N. Leith, J. Upatnieks, Reconstructed wavefronts and communication theory. J. Opt. Soc. Amer. **52**, 1123–1130 (1962)
5. E.N. Leith, J. Upatnieks, Wavefronts reconstruction with continuous-tone objects. J. Opt. Soc. Amer. **53**, 1377–1381 (1963)
6. E.N. Leith, J. Uptnieks, Wavefront reconstruction with diffused illumination and three-dimensional objects. J. Opt. Soc. Amer. **54**, 1295–1301 (1964)
7. E.N. Leith, J. Uptnieks, Holograms: their properties and uses. J. Soc. Photo-Optical Instrum. Engrs J. **4**, 3–6 (1965)
8. E.N. Leith, P. Naulleau, D. Dilworth, Ensemble-averaged imaging through highly scattering media. Opt. Lett. **21**, 1691–1693 (1996)
9. M. Francon, *Holography* (Masson, Paris, 1966)
10. P. Hariharan, *Optical Holography* (Cambridge University Press, Cambridge, 1996). ISBN 0-521-43965-5
11. P. Hariharan, *Basics of Holography* (Cambridge University Press, Cambridge, 2002). ISBN 0-521-00200-1
12. A. VanderLugt, Signal detection by complex spatial filtering. IEEE Trans. Inform. Theory IT **10**, 139–145 (1964)
13. A.M. Hamed, Recognition of colored objects using thick holographic multiplexed filter (THMF). Optica Applicate **13**, 205–213 (1983)
14. J.W. Goodman, *Introduction to Fourier Optics* (McGraw Hill Series, 1996)
15. A.M. Hamed, J. Fleuret, E. Umdenstock, Pattern analysis in polychromatic light, in *International Conference on Image Analysis and Processing*, Pavia Italy (1980)

Chapter 11
Operator Algebra in Complex Coherent Optical Systems

More complex optical systems can be analyzed by using the same methods applied to the optical systems of a single thin lens. However, the number of integrations grows as the number of free-space regions grows, and the complexity of the calculations increases as the number of lenses included grows. For these reasons, some readers may appreciate the introduction of a certain "operator" notion that is useful in analyzing complex systems. Not all readers will find this approach attractive, and for those, the methods already used can simply be extended to more complex systems.

The operator approach is based on several fundamental operations [1, 2], each of which is represented by an "operator".

11.1 Introduction

Early, the impulse response or the Point Spread Function (PSF) is computed for circular and annular apertures in many articles [3–6] using uniform illumination. The conventional microscope resolution is based on the resolution limit of the solely objective lens while the resolution is improved for the confocal scanning laser microscope [7–10] since the two microscope objectives play equal roles in the resolution and the effective PSF is the multiplication of the PSF corresponding to each objective lens [11–15]. Later, the resolution is further improved by investigating the PSF in the case of conic and linear [16, 17], quadratic, and higher-order apertures [18]. In addition, twofold symmetry and fourfold symmetry apertures [19] are proposed, and the PSF is computed and compared with a circular aperture. Recently, different modulated apertures have been suggested, e.g., elliptical apertures and some deformed apertures [20], graded index apertures [21], and the PSF is computed in all cases. In addition, Hamming [22], linear-quadratic [23], and longitudinal strips [24] are investigated. The computations of the PSF corresponding to all the modulated apertures assume coherent uniform illumination.

A. Hamed, *Holographic Imaging Using Aperture Modulation*,
SpringerBriefs in Applied Sciences and Technology,
https://doi.org/10.1007/978-3-031-96989-8_11

The operator approach is based on several fundamental operations [1, 2], each of which is represented by an "operator". Most operators have parameters that depend on the geometry of the optical system being analyzed. Parameters are included within square brackets following the operator. The operators act on the quantities contained in curly brackets.

In this chapter, we start with the basics of operator algebra. Then, the above studies of the modulated apertures for the computation of the PSF of a microscope objective are made for non-uniform Gaussian illumination. The PSF is computed from the Fourier transform of the multiplication of the Gaussian function and the aperture function using operator algebra. The result of the PSF computed from the convolution product of the PSF in case of uniform aperture with Fourier spectrum of the Gaussian function. These results are compared with the corresponding ones obtained in the case of uniform illumination for all the considered modulated apertures. Finally, the results are discussed, and a conclusion is given. The basic operators of use to us here are as follows:

11.2　Basic Operators

1. **Multiplication by a Quadratic-Phase Exponential**. The definition of operator Q is

$$Q[c]\{U(x)\} = e^{j\frac{k}{2}cx^2}U(x), \tag{11.1}$$

   where $k = 2\pi/\lambda$ and c is an inverse length. The inverse of $Q[c]$ is $Q[-c]$.
2. **Scaling by a Constant**. This operator is represented by the symbol v and is defined by

$$v[b]\{U(x)\} = b^{1/2}U(x), \tag{11.2}$$

   where b is dimensionless. The inverse of $v[b]$ is $v[1/\mathbf{b}]$.
3. **Fourier Transformation**. This operator is represented by the usual symbol F and is defined by

$$F\{U(x)\} = \int_{-\infty}^{\infty} U(x)e^{-j2\pi fx}\mathrm{d}x. \tag{11.3}$$

The inverse Fourier transform operator is defined in the usual way, i.e., with a change of the sign of the exponent.

4. **Free-Space Propagation**. Free-space propagation is represented by the operator R, which is defined by the equation:

$$R[d]\{U(x_1)\} = \frac{1}{\sqrt{j\lambda d}} \int\limits_{-\infty}^{\infty} U(x_1) e^{j\frac{k}{2d}(x_2-x_1)^2} \, dx_1, \tag{11.4}$$

where d is the distance of propagation and x_2 is the coordinate that applies after propagation.

The inverse of $R[d]$ is $R[-d]$.

These four operators are sufficient for analyzing most optical systems. Their utility arises from some simple properties and certain relations between them. These properties and relations allow complicated chains of operators to be reduced to simple results, as will shortly be illustrated. Some simple and useful properties are listed below:

$$v[t_2]v[t_1] = v[t_2t_1], \tag{11.5}$$

$$Fv[t] = v\left[\frac{1}{t}\right]F, \tag{11.6}$$

$$FF = v[-1], \tag{11.7}$$

$$Q[c_2]Q[c_1] = Q[c_2 + c_1], \tag{11.8}$$

$$R[d] = F^{-1}Q\left[-\lambda^2 d\right]F, \tag{11.9}$$

$$Q[c]v[t] = v[t]Q\left[\frac{c}{t^2}\right]. \tag{11.10}$$

Relations (11.5) and (11.8) are obvious and simple to prove. Relation (11.6) is a statement of the similarity theorem of Fourier analysis, while Relation (11.7) follows from the Fourier inversion theorem, slightly modified to account for the fact that both transforms are in the forward direction. Relation (11.9) is a statement that free-space propagation over distance can be analyzed either by a Fresnel diffraction equation or by a sequence of Fourier transformation, multiplication by the transfer function of free space, and inverse Fourier transformation. The left-hand and right-hand sides of Relation (11.10) are shown to be equal simply by writing out their definitions. A slightly more sophisticated relation is

$$R[d] = Q\left[\frac{1}{d}\right]v\left[\frac{1}{\lambda d}\right]FQ\left[\frac{1}{d}\right] \tag{11.11}$$

which is a statement that the Fresnel diffraction operation is equivalent to pre-multiplication by a quadratic-phase exponential, a properly scaled Fourier transform, and post-multiplication by a quadratic-phase exponential. Another relation of similar complexity is

$$v\left[\frac{1}{\lambda f}\right]F = F[f]Q\left[-\frac{1}{f}\right]R[f] \qquad (11.12)$$

which is a statement that the fields across the front and back focal planes of a positive lens are related by a properly scaled Fourier transform, with no quadratic-phase exponential multiplier, as proved earlier in this chapter.

Many useful relations between operators are summarized in Table 11.1. With these relations to draw on, we are now ready to apply the operator notation to some simple optical systems.

11.3   Computation of the PSF Using Operator Algebra in a Gaussian Beam Illumination

Assuming the complex amplitude of the Gaussian beam emitted from the laser is represented as follows:

$$g(\rho;\ z) = g_0 \exp\left\{j\left[p(z) + \frac{k\rho^2}{2q(z)}\right]\right\}, \qquad (11.13)$$

where ρ is the spatial plane of coordinates (x, y), and z is the longitudinal or axial coordinate for the beam propagation. $P(z)$ is a phase factor depending on z, and $q(z)$ is the complex radius of curvature of the beam. Q is defined by the radius of curvature of the beam (R) and the width of the beam (w) as follows:

$$\frac{1}{q} = \frac{1}{R} + j\frac{\lambda}{\pi w^2}. \qquad (11.14)$$

Substituting from (11.14) in (11.13), we can write the transverse Gaussian beam in the form:

$$g(\rho) = A \exp\left(j\frac{k\rho^2}{2R}\right)\exp\left(-\frac{\rho^2}{w^2}\right), \qquad (11.15)$$

where A contains the longitudinal phase factor that has no transversal dependence.

The pupil is situated near the converging lens L where the distance between them is neglected. The incident Gaussian beam of complex amplitude $u_1(\rho)$ is represented by operators [1, 2] as follows:

$$u_1(\rho) = Q\left[\frac{1}{R}\right]Q\left[j\frac{\lambda}{\pi w^2}\right]A = Q\left[\frac{1}{q}\right]A. \qquad (11.16)$$

Table 11.1 Relations between operators [1]

	v	F	Q	R
v	$v[t_2]v[t_1] = v[t_2t_1]$	$v[t]F = Fv\left[\frac{1}{t}\right]$	$v[t]Q[c] = Q[t^2]v[t]$	$v[t]R[d] = R\left[\frac{d}{t^2}\right]v[t]$
F	$Fv[t] = v\left[\frac{1}{t}\right]F$	$FF = v[-1]$	$FQ[c] = R\left[-\frac{c}{\lambda^2}\right]F$	$FR[d] = Q[-\lambda^2 d]F$
Q	$Q[c]v[t] = v[t]Q\left[\frac{c}{t^2}\right]$	$Q[c]F = FR\left[-\frac{c}{\lambda^2}\right]$	$Q[c_2]Q[c_1] = Q[c_2 + c_1]$	$Q[c]R[d] = R\left[(d^{-1} + c)^{-1}\right] \cdot v[(1 + cd)] \cdot Q\left[(c^{-1} + d)^{-1}\right]$
R	$R[d]v[t] = v(t)R[t^2 d]$	$R[d]F = FQ[-\lambda^2 d]$	$R[d]Q[c] = Q\left[(c^{-1} + d)^{-1}\right] \cdot v\left[(1 + cd)^{-1}\right] \cdot R\left[(d^{-1} + c)^{-1}\right]$	$R[d_2]R[d_1] = R[d_1 + [d_2]]$

W is the beam width in the plane (x, y), while $r = (u, v)$ is the radial coordinate in the Fourier plane.

In Eq. (11.16), the quadratic operator is defined as: $Q[a] = e^{\frac{jka\rho^2}{2}}$; $a = \frac{1}{q}$.

The transfer operator is mainly affected by the shape of the aperture and the lens L. The complex amplitude in the plane at the focal plane of the lens $u_2(r)$ is related to the input Gaussian beam $u_1(\rho)$ through the transfer operator τ as:

$$u_2(r) = \tau u_1(\rho). \tag{11.17}$$

Making use of the optical system shown in Fig. 11.1, the transfer operator is computed as follows:

$$\tau = R[f]L[f]P(\rho) = R[f]Q\left[-\frac{1}{f}\right]P(\rho). \tag{11.18}$$

Substitute Eqs. (11.16, 11.18) in Eq. (11.17), we write:

$$u_2(r) = R[f]Q\left[-\frac{1}{f}\right]P(\rho)Q\left[\frac{1}{R}\right]Q\left[j\frac{\lambda}{\pi w^2}\right]A. \tag{11.19}$$

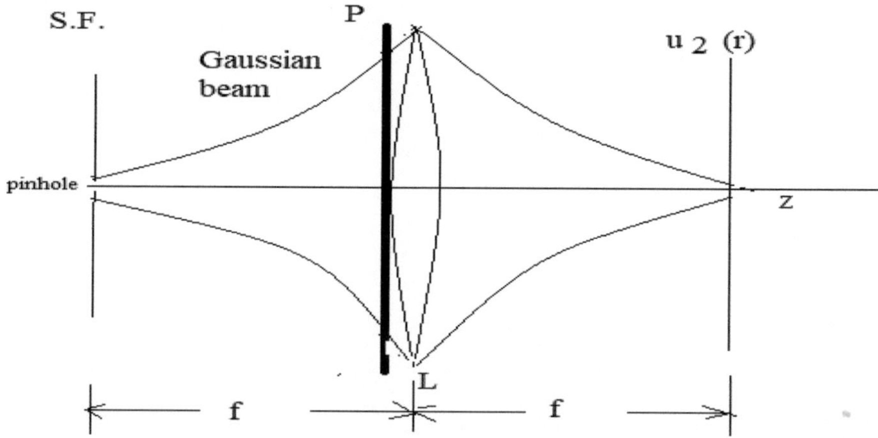

Fig. 11.1 Non-uniform Gaussian beam is incident upon an aperture $P(\rho)$ followed by a converging lens of focal length f. The PSF is in the focal plane of the lens L. S.F. is a spatial filter

Assume that the beam radius of curvature $R = f$ is the focal length of the converging lens. In this case, the quadratic terms in Eq. (11.19) are grouped to be unity since $Q[a]Q[-a] = Q[0] = 1$. Then, Eq. (11.19) becomes:

$$u_2 = AR[f]Q\left[j\frac{\lambda}{\pi w^2}\right]P(\rho).\tag{11.20}$$

Making use of the Fresnel Free-Space Propagation (FPO) represented as follows [2]:

$$R[z] = \frac{e^{jkz}}{j\lambda z}Q\left[\frac{1}{z}\right]v\left[\frac{1}{\lambda z}\right]FQ\left[\frac{1}{z}\right].\tag{11.21}$$

Then, Eq. (11.20) becomes:

$$u_2 = A\frac{e^{jkf}}{j\lambda f}Q\left[\frac{1}{f}\right]v\left[\frac{1}{\lambda f}\right]FQ\left[\frac{1}{f}\right]Q\left[j\frac{\lambda}{\pi w^2}\right]P(\rho).\tag{11.22}$$

Equation (11.22) is rewritten as follows:

$$u_2 = A\frac{e^{jkf}}{j\lambda f}Q\left[\frac{1}{f}\right]v\left[\frac{1}{\lambda f}\right]FQ\left[\frac{1}{f}+j\frac{\lambda}{\pi w^2}\right]P(\rho).\tag{11.23}$$

Let $\frac{1}{f}+j\frac{\lambda}{\pi w^2} = a$, and using the following property:

$$(FQ[a]P(\rho)) = \{(FQ[a])\otimes(FP(\rho))\} = \left\{\frac{j\lambda}{a}Q\left[-\frac{\lambda^2}{a}\right]\otimes(FP(\rho))\right\}.\tag{11.24}$$

Substitute from (11.24) in (11.23), we get finally:

$$u_2 = A\frac{e^{jkf}}{1+\left(j\frac{\lambda f}{\pi w^2}\right)}Q\left[\frac{1}{f}\right]v\left[\frac{1}{\lambda f}\right]Q\left[-\frac{\lambda^2}{\frac{1}{f}+j\frac{\lambda}{\pi w^2}}\right]\otimes(FP(\rho)).\tag{11.25}$$

Since the Fourier transform of the pupil function gives the PSF in case of uniform illumination, hence Eq. (11.25) is rewritten as follows:

$$u_2(r) = A\frac{e^{jkf}}{1+\left(j\frac{\lambda f}{\pi w^2}\right)}Q\left[\frac{1}{f}\right]v\left[\frac{1}{\lambda f}\right]Q\left[-\frac{\lambda^2}{\frac{1}{f}+j\frac{\lambda}{\pi w^2}}\right]\otimes(h(r)),\tag{11.26}$$

where $h(r) = \text{F.T.}\,[P(\rho)]$.

Consequently, the complex amplitude in the focal plane of the lens L is the convolution of the phase-distorted PSF $= h(r)$, for the uniform illumination with the Gaussian amplitude distribution.

The intensity of the image of a point in the case of Gaussian illumination is computed by taking the modulus square of Eq. (11.26):

$$I(r) = \left| v\left[\frac{1}{\lambda f}\right] Q\left[-\frac{\lambda^2}{\frac{1}{f} + j\frac{\lambda}{\pi w^2}}\right] \otimes (h(r)) \right|^2. \tag{11.27}$$

In the case of uniform illumination instead of Gaussian beam, the convolution of the PSF with Dirac-Delta function gives the PSF in case of uniform illumination. Hence, we get the image of a point object as:

$$I(w) = \left| v\left[\frac{1}{\lambda f}\right](h(r)) \right|^2 = |h(w)|^2 \tag{11.28}$$

$w = \rho_0 \frac{r}{\lambda f}$ is the reduced coordinate in the Fourier plane. ρ_0 is the aperture radius.

Table 11.2 summarizes the results of the PSF obtained by the author in [16–23] assuming uniform illumination. These results are convoluted with the Gaussian function in Eq. (11.26) to get the PSF in the case of Gaussian illumination. Consequently, the image of a point or the intensity impulse response is computed from Eq. (11.27) for the Gaussian beam.

Table 11.2 PSF in case of uniform illumination for some modulated apertures

Aperture $P(\rho)$	PSF in case of uniform illumination
1. Circular	$h(w) = 2\,J_1(w)/w \to$ Airy disk; $w = r\,(\rho_0/\lambda f)$
2. Annular	$h(w) = J_0(w)$
3. B/W concentric annuli	$h(w) = \sum\limits_{n=1}^{N} \left\{ r_n'^2\left[\frac{2J_1(w_n')}{w_n'}\right] - r_n^2\left[\frac{2J_1(w_n)}{w_n}\right]\right\}$ [17]
4. Linear	$h(w) = 4\pi\left[\frac{J_1(W)}{W} + \frac{J_0(W)}{W^2} - \frac{2\sum_i J_i(W)}{W^3}\right]$ [17]
5. Quadratic	$h(w) = \text{const.}\left[\frac{J_1(W)}{W} - \frac{J_2(W)}{W^2}\right]$ [18]
6. Twofold symmetrical	A. M. Hamed, Optics and laser tech. 1984 [19]
7. Fourfold symmetry	A.M. Hamed, Optics and laser 1984 [19]
8. Linear-quadratic	A. M. Hamed, Optik 2017 [23]
9. Hamming aperture	$h(r) =$ $\left\{\delta(r) + \left(\frac{1}{2}\right)\left[\delta_1\left(r - \beta\lambda\frac{f}{2} - \beta\lambda\rho_2\right) + \delta_2\left(r + \beta\lambda\frac{f}{2} + \beta\lambda\rho_2\right)\right]\right\}$ [22]
10. Longitudinal strips inside the circular aperture	$. \, h(r) = 2\frac{J_1(\alpha r)}{\alpha r} \otimes u_0 \frac{\sin\left(\pi\frac{u_0 x}{\lambda f}\right)}{\pi\frac{u_0 x}{\lambda f}} \sum\limits_{n=0}^{N} \exp\left[-j2\pi(2n+1)u_d x/\lambda f\right]$ [24]

The quadratic term multiplied from the left by the scaling factor in Eq. (11.26) is decomposed into a radial quadratic-phase term and the modulated Gaussian beam as follows:

$$v\left[\frac{1}{\lambda f}\right]Q\left[-\frac{\lambda^2}{\frac{1}{f}+j\frac{\lambda}{\pi w^2}}\right] = Q\left[(1/\lambda f)^2\left(-\frac{\lambda^2}{\frac{1}{f}+j\frac{\lambda}{\pi w^2}}\right)\right] = Q\left[-\frac{1}{f+j\frac{\lambda f^2}{\pi w^2}}\right].$$

$$\tag{11.29}$$

After separating the real and imaginary parts, we finally get:

$$Q\left[-\frac{1}{f+j\frac{\lambda f^2}{\pi w^2}}\right] = Q\left[-\frac{f}{f^2+\left(\frac{\lambda^2 f^4}{\pi^2 w^4}\right)}\right]Q\left[\frac{j\frac{\lambda f^2}{\pi w^2}}{f^2+\left(\frac{\lambda^2 f^4}{\pi^2 w^4}\right)}\right]. \tag{11.30}$$

In exponential form, the quadratic terms in Eq. (11.30) are written as follows: The quadratic radial phase term is:

$$Q\left[-\frac{f}{f^2+\left(\frac{\lambda^2 f^4}{\pi^2 w^4}\right)}\right] = \exp\left\{-\frac{j\pi\left(\frac{f}{\lambda}\right)r^2}{\left[f^2+\left(\frac{\lambda^2 f^4}{\pi^2 w^4}\right)\right]}\right\}. \tag{11.31}$$

And the resulting beam is again Gaussian but affected by a weighting factor in the Fourier plane giving this commode distribution as expected.

$$Q\left[\frac{j\frac{\lambda f^2}{\pi w^2}}{f^2+\left(\frac{\lambda^2 f^4}{\pi^2 w^4}\right)}\right] = \exp\left\{-\frac{r^2/w^2}{\left[1+\left(\frac{\lambda^2 f^2}{\pi^2 w^4}\right)\right]}\right\} = \exp\{-r^2/(\alpha w^2)\}, \tag{11.32}$$

where $\alpha = 1+\left(\frac{\lambda^2 f^2}{\pi^2 w^4}\right)$ is the weighting factor.

From Eqs. (11.26) and (11.32), we compute PSF in the case of Gaussian propagation.

The results presented in Table 11.2 for the PSF in the case of uniform illumination are convoluted with the Fourier spectrum of the Gaussian illumination function. Hence, we obtain the PSF in the case of Gaussian illumination. It is known that the Fourier spectrum of the Gaussian illumination function is computed to give a Gaussian function (Figs. 11.2 and 11.3).

$$\text{F. T.}\exp\left\{-\frac{\rho^2}{w^2}\right\} = \exp\left\{-\frac{w^2 r^2}{\lambda^2 f^2}\right\}. \tag{11.33}$$

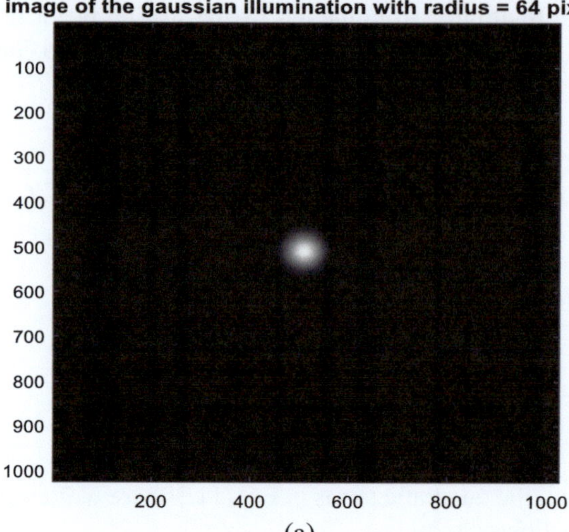

(a)

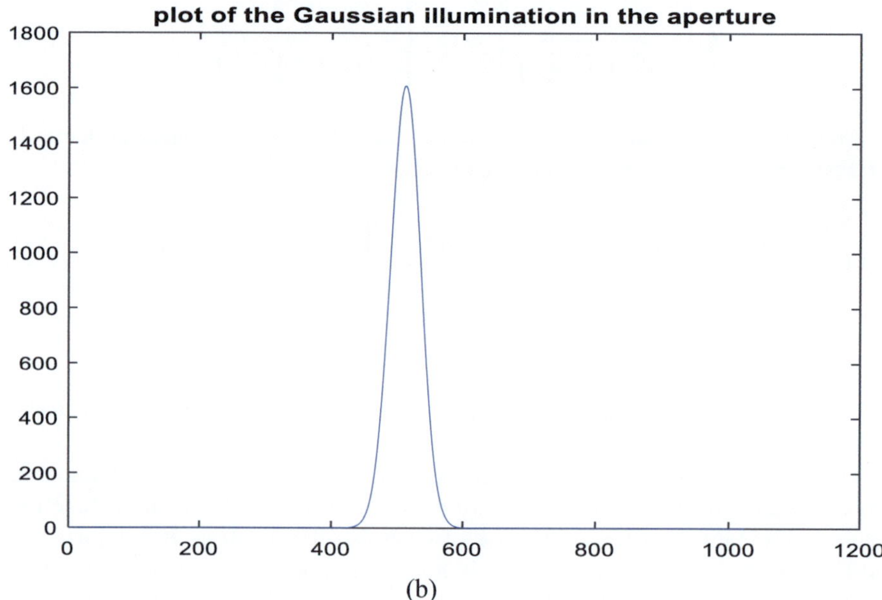

(b)

Fig. 11.2 **a** Image of the illumination Gaussian beam. The aperture radius $R = 64$ pixels. **b** Line plot of the Gaussian beam. The truncation radius $= 64$ pixels. **c** F.T. of the Gaussian beam

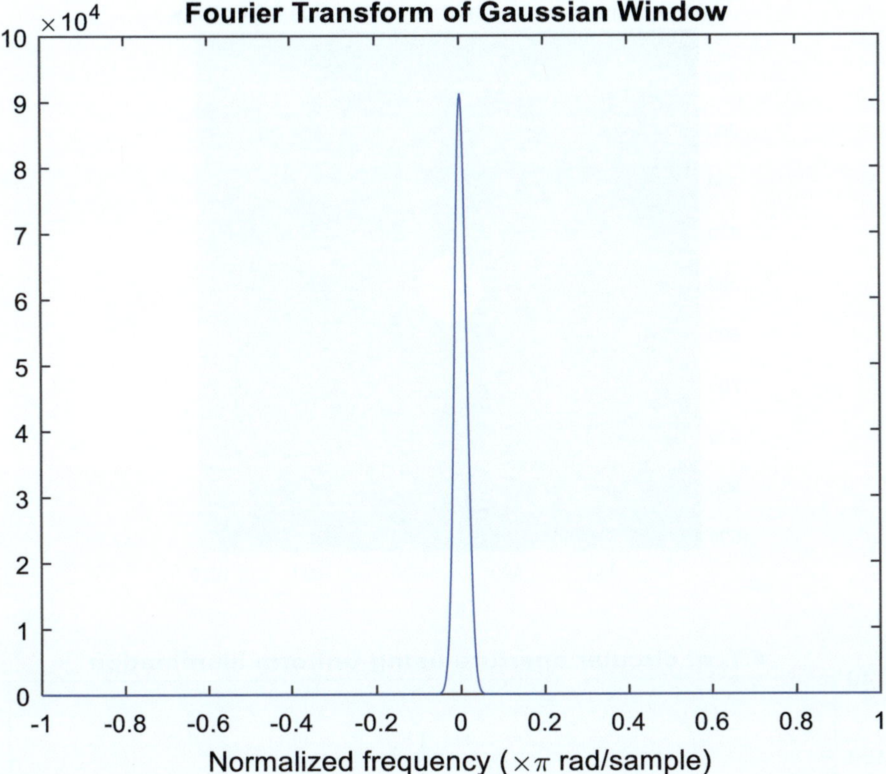

Fig. 11.2 (continued)

11.4 Results and Discussion

The PSF for the Gaussian illumination assuming circular aperture computed from the FFT is plotted for different radii. These results are plotted as in Fig. 11.4.

The resulting PSF where $\sigma = 32$ pixels and the radius for the circular aperture $= 64$ pixels is represented as in Fig. 11.4a. It showed truncation of the legs in the PSF pattern. The resulting PSF where $\sigma = 64$ pixels and the radius for the circular aperture $= 64$ pixels is represented as in Fig. 11.4b.

The PSF where $\sigma = 32$ pixels for the Gaussian beam and the radius for the circular aperture $= 8$ pixels is shown in Fig. 11.5a. In Fig. 11.5b, the linear aperture is considered a modulated aperture used for the computation of the PSF in the case of Gaussian illumination where $\sigma = 32$ pixels and the maximum radius $= 8$ pixels for the linear aperture. In addition, results corresponding to the quadratic aperture are plotted in Fig. 11.5c. A comparison between the results represented in Fig. 11.5a–c, and it is shown better resolution in the case of quadratic and linear apertures as compared with circular apertures considering Gaussian illumination of the same

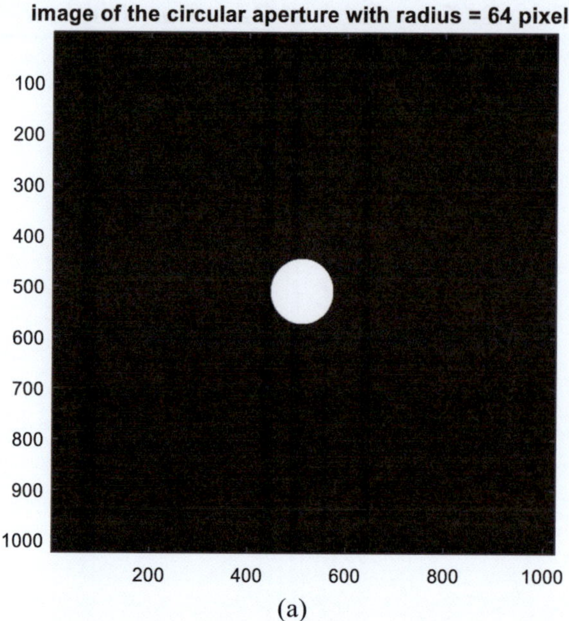

(a)

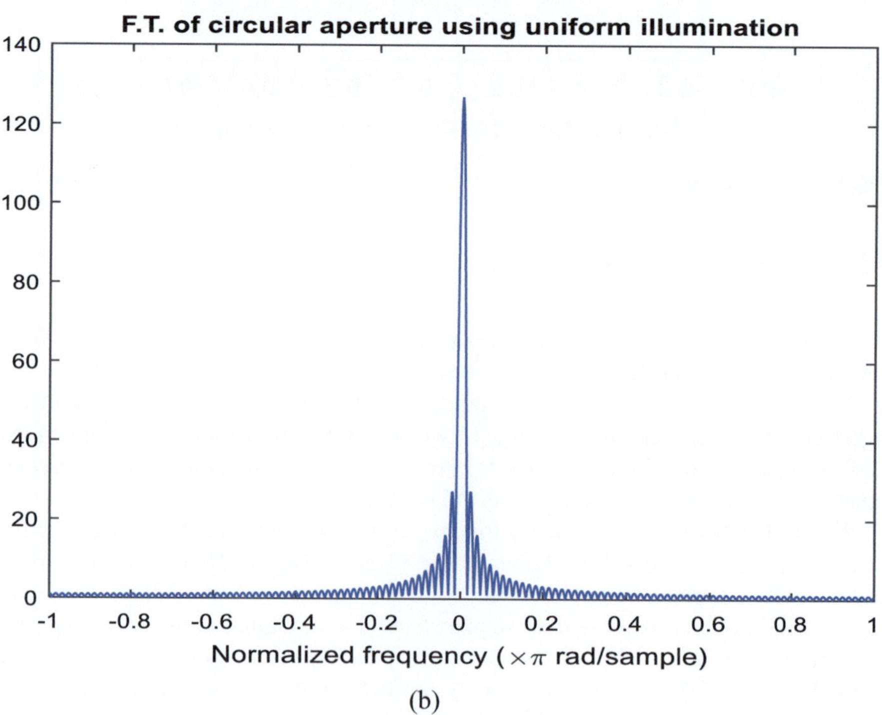

(b)

Fig. 11.3 **a** Circular aperture placed in the converging lens plane. **b** PSF in case of uniform illumination

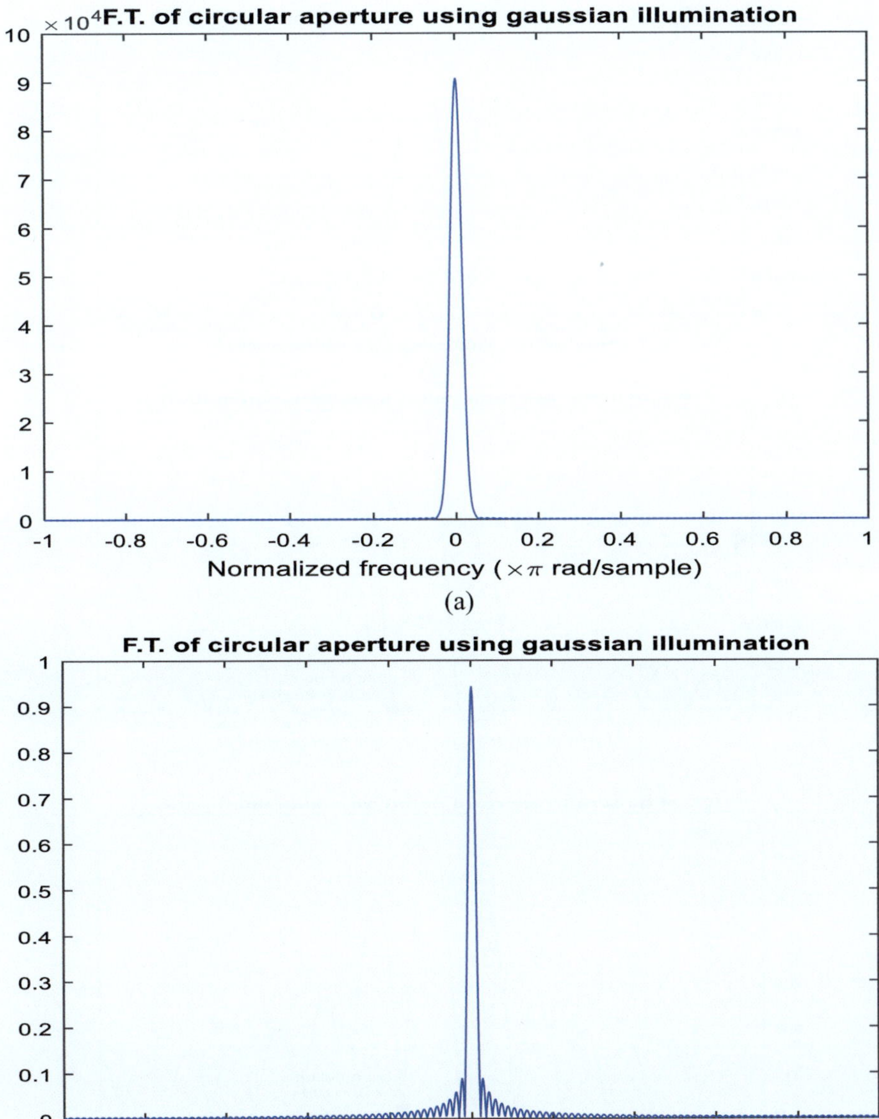

Fig. 11.4 **a** Resulted PSF where $\sigma = 32$ pixels. The radius $= 64$ pixels for the circular aperture. It showed truncation of the legs in the PSF pattern. **b** Resulted PSF where $\sigma = 64$ pixels and the radius $= 64$ pixels for the circular aperture

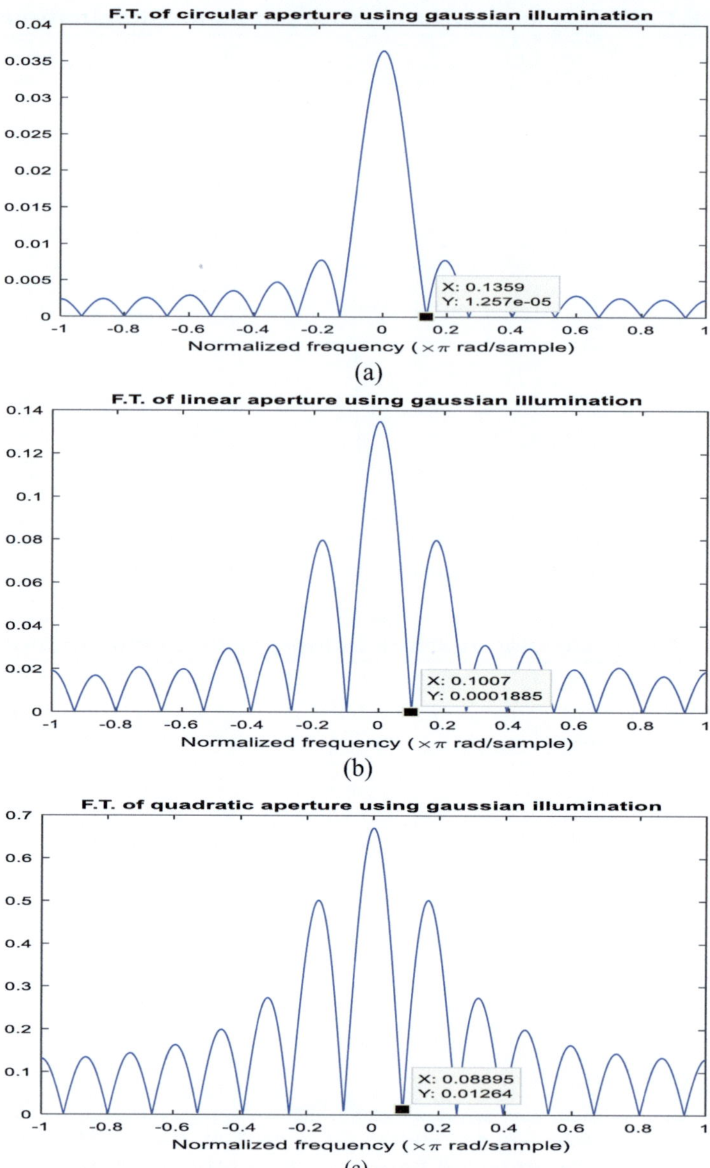

Fig. 11.5 **a** Resulted PSF where $\sigma = 32$ pixels for the Gaussian beam and the radius $= 8$ pixels for the circular aperture. **b**: Resulted PSF where $\sigma = 32$ pixels for the Gaussian beam, and maximum radius $= 8$ pixels for the linear aperture. **c**: Resulted PSF where $\sigma = 32$ pixels for the Gaussian beam, and maximum radius $= 8$ pixels for the quadratic aperture

truncation value σ. The better resolution is governed by the cut-off spatial frequency r_c. It is shown that the cut-off spatial normalized frequency $r_c = 0.1359$ in the case of circular aperture Fig. 11.5a is greater than the normalized frequency $r_c = 0.1007$ in the case of linear aperture as in the Fig. 11.5b. Further improvement is shown for the quadratic aperture giving $r_c = 0.08895$ as shown in Fig. 11.5c. In addition, the legs for the quadratic and linear apertures are higher than that corresponding to the circular aperture considering Gaussian illumination which is useful for imaging the extended objects. Another set of plots of PSF where $\sigma = 64$ pixels and the radius for circular, linear, and quadratic apertures is set constant as before and equals 8 pixels as given in Fig. 11.6a–c. It is shown that the cut-off normalized frequency is given as:

$r_c = 0.1339$(circular) $> r_c = 0.09873$ (linear) $> r_c = 0.09091$(quadratic); Gaussian illumination.

Hence, the quadratic aperture provided with Gaussian illumination has better resolution than the linear aperture followed by the circular aperture.

The PSF for uniform illumination for the circular, linear, and quadratic apertures is represented in Fig. 11.7a–c showing these results for the normalized cut-off frequencies. In this case, we get these results using uniform illumination to be compared using Gaussian illumination:

$r_c = 0.1359$(circular) $> r_c = 0.1026$ (linear) $> r_c = 0.09091$(quadratic); uniform illumination.

The direct convolution of the Airy disk obtained for uniform illumination and the Fourier spectrum corresponding to Gaussian illumination for different truncations of the Gaussian function (σ values) giving the PSF plotted as in Fig. 11.8. The above direct convolution is computed from the following equation:

$$h(r) = \exp\left\{-\frac{r^2}{\sigma^2}\right\} \otimes \frac{\left\{2J_1\left(\frac{2\pi\rho_0 r}{\lambda f}\right)\right\}}{\left(\frac{2\pi\rho_0 r}{\lambda f}\right)}; \quad \text{for circular aperture,} \tag{11.34}$$

where $w = \lambda f/\sigma$ is the laser beam waist.

The results presented in Fig. 11.8 to compute the PSF in the case of Gaussian illumination using direct convolution showed truncation of the legs in the PSF for $\sigma < 1.25\,\pi$, while the legs have appeared as the beam waist $w = \sigma$ is decreased. The results obtained in Fig. 11.8 for circular aperture illuminated with Gaussian beam.

The PSF corresponding to the Gaussian illumination and linear distribution inside the circular aperture is computed from this equation:

$$h(r) = \exp\left\{-\frac{r^2}{\sigma^2}\right\} \otimes \left[\frac{J_1(W)}{W} + \frac{J_0(W)}{W^2} - \frac{2\sum_i J_i(W)}{W^3}\right]; \quad \text{for linear aperture,}$$

$$\tag{11.35}$$

where $W = \left(\frac{2\pi\rho_0 r}{\lambda f}\right)$ is the reduced coordinate in the Fourier plane.

The PSF in the case of quadratic aperture is computed from the following equation:

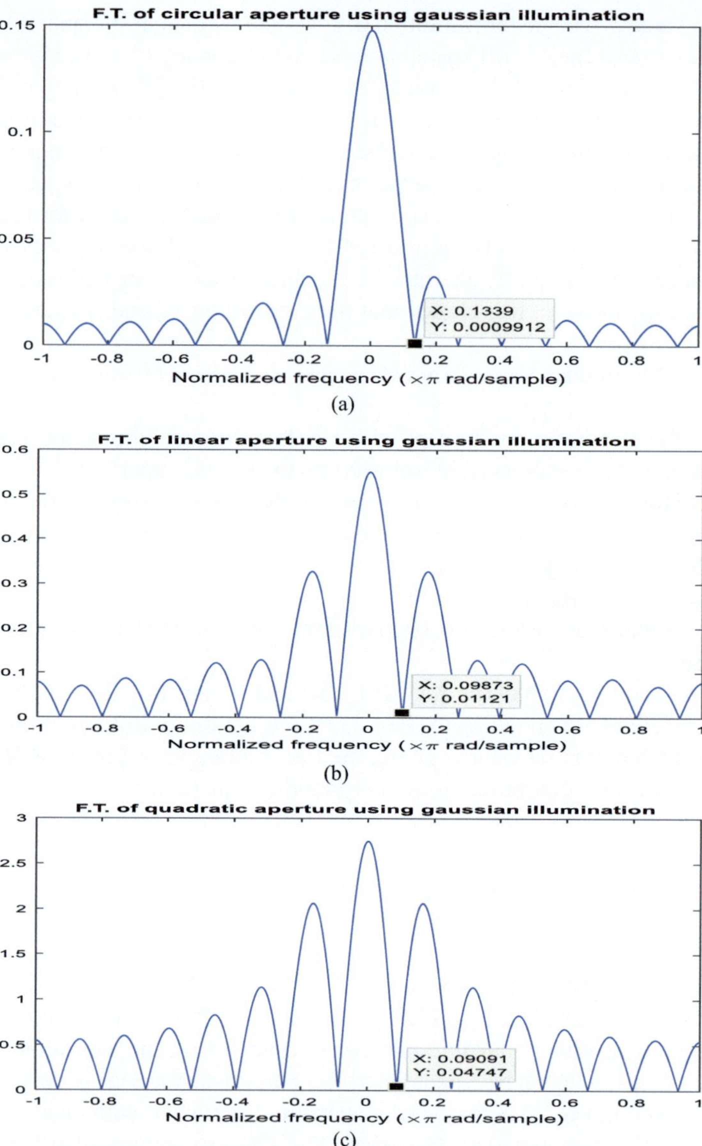

Fig. 11.6 **a** Resulted PSF where $\sigma = 64$ pixels for the Gaussian beam and the radius $= 8$ pixels for the circular aperture. **b** Resulted PSF where $\sigma = 64$ pixels for the Gaussian beam, and maximum radius $= 8$ pixels for the linear aperture. **c** Resulted PSF where $\sigma = 64$ pixels for the Gaussian beam, and maximum radius $= 8$ pixels for the quadratic aperture

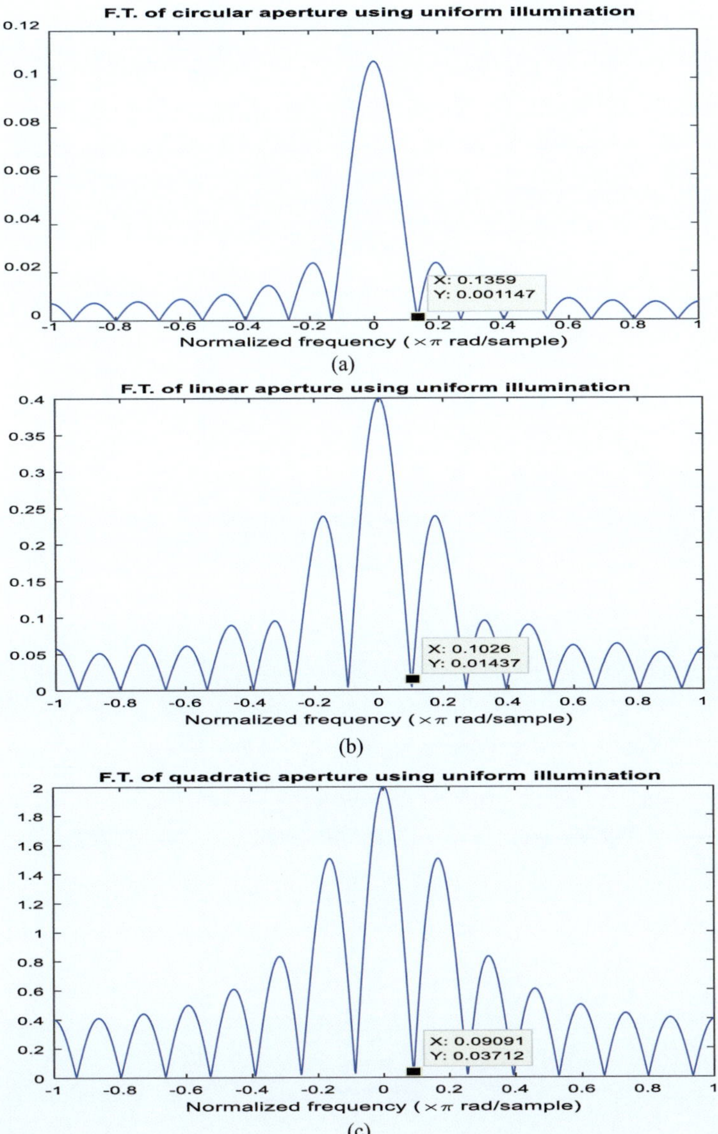

Fig. 11.7 **a** PSF for the uniform circular aperture where the maximum radius = 8 pixels using uniform illumination. **b** PSF for the linear aperture where the maximum radius = 8 pixels using uniform illumination. **c** PSF for the quadratic aperture where the maximum radius = 8 pixels using uniform illumination

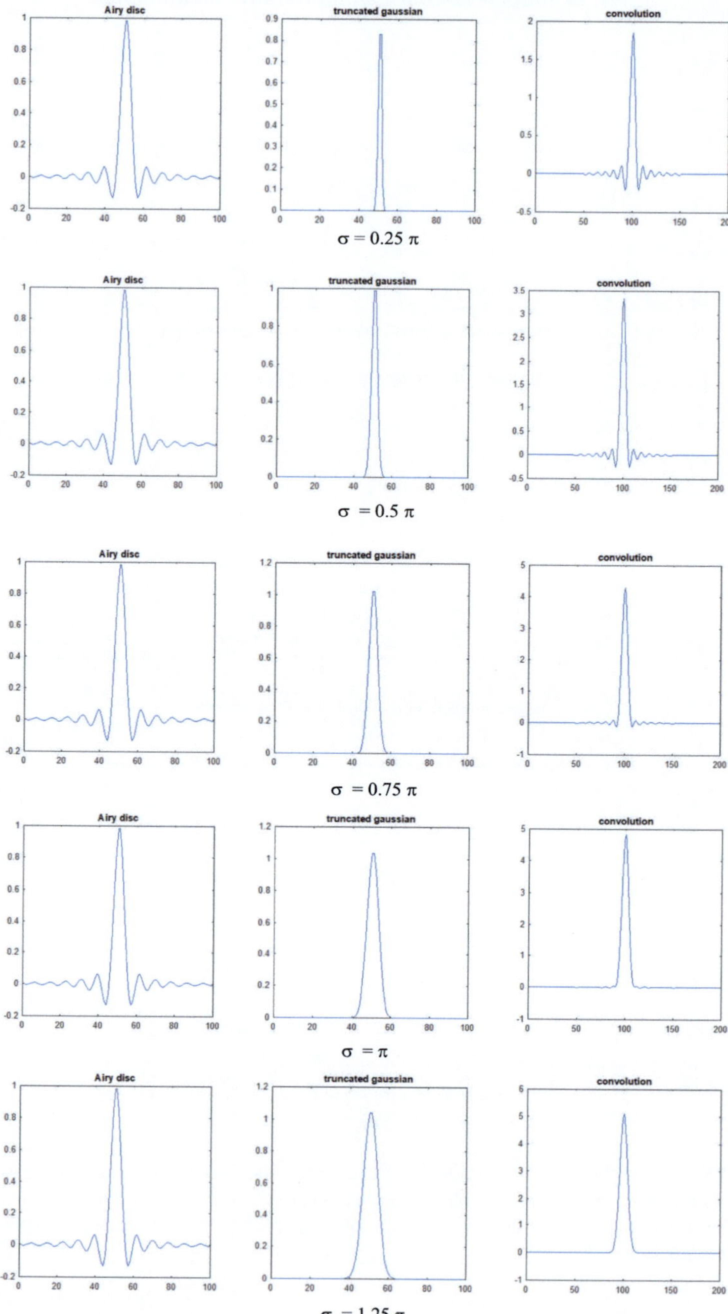

Fig. 11.8 PSF in case of Gaussian illumination. It is computed from the direct convolution product of the PSF in the case of uniform illumination (Airy disk) and the Fourier spectrum of the Gaussian illumination beam. Different values of σ in Eq. (11.22) showed truncation of the legs in the PSF for $\sigma < 1.25\,\pi$

$$h(r) = \exp\left\{-\frac{r^2}{\sigma^2}\right\} \otimes \left[\frac{J_1(W)}{W} - \frac{J_2(W)}{W^2}\right]; \quad \text{for quadratic aperture.} \qquad (11.36)$$

The PSF in case of annular aperture illuminated with the Gaussian beam:

$$h(r) = \exp\left\{-\frac{r^2}{\sigma^2}\right\} \otimes J_0(W); \quad \text{for annular aperture.} \qquad (11.37)$$

The results of the PSF using Gaussian illumination for quadratic aperture and comparison with the case of circular aperture are plotted as shown in Fig. 11.9. $C1$ stands for the direct convolution of the Airy disk and the Fourier spectrum of the Gaussian function, while $C2$ stands for the direct convolution of the PSF corresponding to the quadratic aperture and the Fourier spectrum of the Gaussian function. Five values of σ are given ranging from 0.25π up to 1.25π. The legs truncation is shown for $\sigma \geq \pi$.

11.5 Conclusion

It is demonstrated that operator algebra is useful for investigating the PSF by utilizing Fourier optics and convolution operations. The application of two famous modulation methods, namely, linear and quadratic apertures, placed in front of the converging lens using Gaussian illumination, is presented. A comparison with the circular aperture is made and shows important results. The PSF shape remains unchanged with Gaussian illumination and the resolution is improved in the case of linear and quadratic apertures since the normalized cut-off spatial frequency is diminished for the same aperture radius and truncation parameter σ. In addition, the dominance of secondary peaks compared with circular aperture and uniform illumination is apparent allowing to imaging of extended objects.

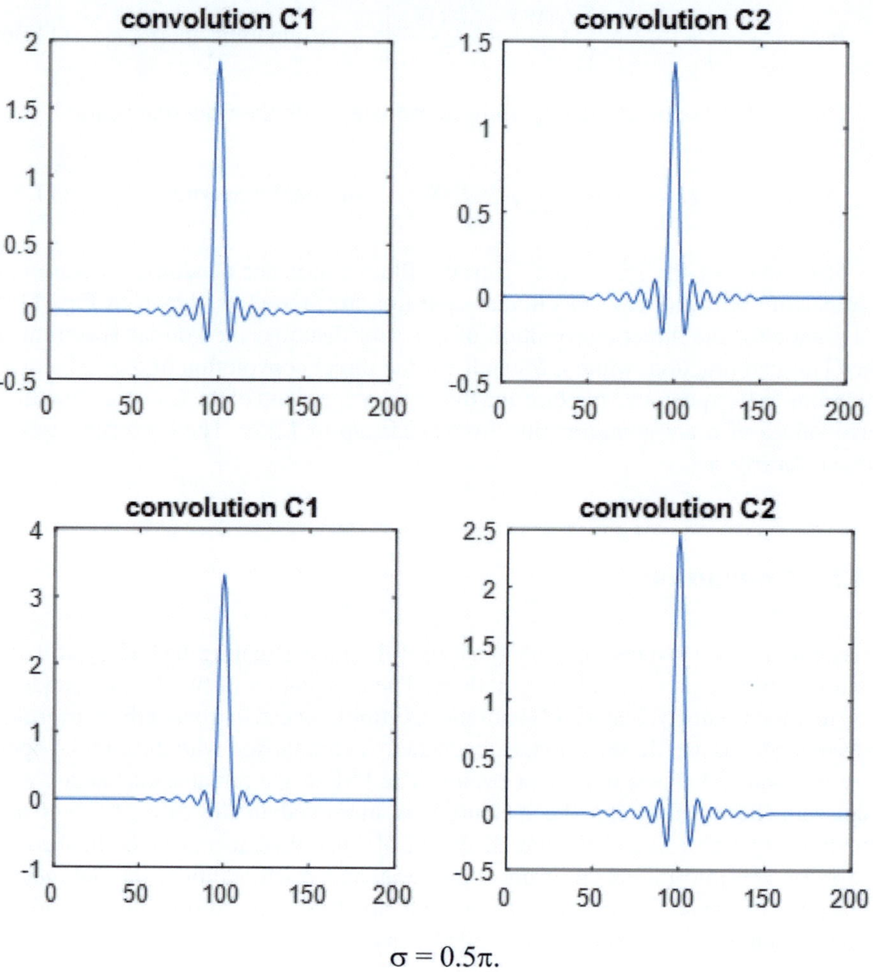

Fig. 11.9 **a–e** $C1$ stands for the direct convolution of the Airy disk and the Fourier spectrum of the Gaussian function, while $C2$ stands for the direct convolution of the PSF corresponding to the quadratic aperture and the Fourier spectrum of the Gaussian function. $\sigma = 0.25\pi$, 0.5π, 0.75π, π, and 1.25π

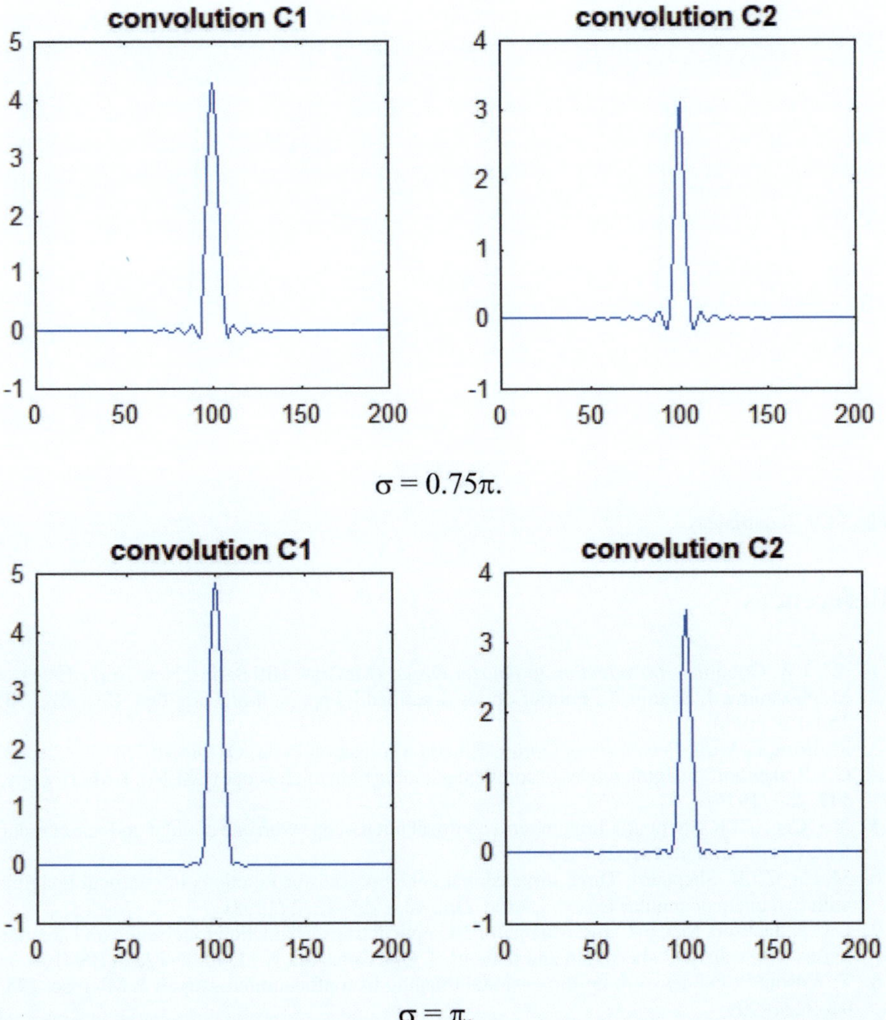

Fig. 11.9 (continued)

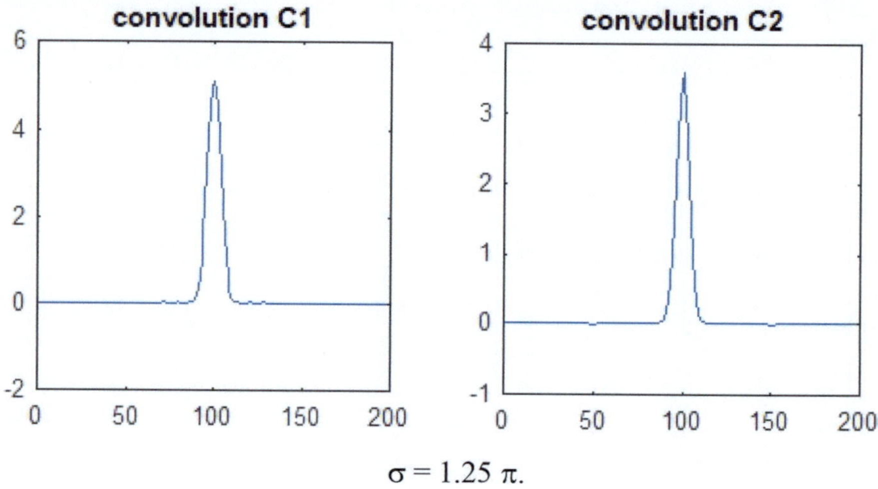

$$\sigma = 1.25\ \pi.$$

Fig. 11.9 (continued)

References

1. .23. J.W. Goodman, *Introduction to Fourier Optics* (McGraw Hill Series, New York, 1996)
2. M. Nazarathy, J. Shamir, J., Fourier optics described by operator algebra. Opt. Soc. Am. **70**, 150–159 (1980)
3. M. Born, E. Wolf, *Principles of Optics*, 5th edn. (Pergamon Press, Oxford, 1975)
4. G.J. Brakenhoff, Imaging modes in confocal scanning light microscope (CSLM). J. Microscopy, **117**, 233 (1979)
5. X.S. Can, C.J.R. Sheppard, Imaging in a confocal microscope with one circular and one annular lens. Opt. Commun. **103**, 254–264 (1993)
6. M. Cu, C.J.R. Sheppard, Three-dimensional coherent transfer functions in confocal imaging with two unequal annular lenses. J. Mod. Opt. **40**, 1255–1272 (1993)
7. D.C.A. Jackson, M. Cu, C.J.R. Sheppard, 3-D optical transfer function for circular and annular lenses with spherical aberration and defocus. J. Opt. Soc. Am. A **11**, 1758–1767 (1994)
8. T. Wilson, Principles of three-dimensional imaging in confocal microscopes. J. Microsc. **193**, 91–92 (1999)
9. T.R. Corle, G.S. Kino, *Confocal Scanning Optical Microscopy and Related Imaging Systems* (Springer, New York, 1996)
10. D. Semwogerere, E.R. Weeks, Confocal microscopy. Biomed. Eng. **23**(2), 1–10 (2005)
11. C.J.R. Sheppard, A. Choudhury, Image formation in the scanning microscope. Opt. Acta **24**, 1051 (1977)
12. T. Wilson, A.R. Carlini, Three-dimensional imaging in confocal imaging systems with finite-sized detectors. J. Microscopy **141**, 51–66 (1988)
13. G. Cox, C.J.R. Sheppard, Practical limits of resolution in confocal and non-linear microscopy. Microscopic Res Tech. **63**, 18–22 (2004)
14. C.J.R. Sheppard, D.M. Shotton, *Image Formation in the Confocal Laser Scanning Microscope* (Springer, New York, 1997), pp. 15–31
15. C.J.R. Sheppard, A. Choudhury, Annular pupils, radial polarization, and super-resolution. Appl. Opt. **43**, 4322–4327 (2004)
16. J.J. Clair, A.M. Hamed, Theoretical studies on optical coherent microscope. Optik **64**, 33–141
17. A.M. Hamed, J. J. Clair, Image and super-resolution in optical coherent microscopes. Optik **64**, 277–284

18. A.M. Hamed, J.J. Clair, Studies on optical properties of confocal scanning optical microscope using pupils with radially transmission distribution. Optik **65**, 209–218 (1983)
19. A.M. Hamed, Resolution and contrast in confocal optical scanning microscope. Opt. Laser Technol **16**, 93–96 (1984)
20. A. M. Hamed, Journal of Optical Engineering 50 (2011) 1–7, Discrimination between speckled images using diffusers modulated by some deformed apertures: Simulations. https://doi.org/10.1117/1.3530085
21. A.M. Hamed, Study of graded index and truncated apertures using speckle images. Precision instrument and mech. PIM **3**, 144–152 (2014)
22. A.M. Hamed, T.A. Al-Saeed, Image analysis of modified Hamming aperture: application on confocal microscopy and holography. J. Modern Opt. **62**, 801–810 (2015). https://doi.org/10.1080/09500340.2015.1007102
23. A.M. Hamed, Improvement of point spread function (PSF) using linear-quadratic aperture. Optik **131**, 838–849 (2017). https://doi.org/10.1016/j.ijleo.2016.11.201
24. A.M. Hamed, *The Point Spread Function for Some Modulated Apertures Application on Speckle and Interferometry Images* (Lambert Academic Publishing (LAP), 2017). ISBN: 978-620-2-07070-6